FORTRAN Computer Programs

FORTRAN Computer Programs

Solutions to Optimization Problems Arising in Feedback Control

C.W. Merriam III
The University of Rochester

Lexington Books
D.C. Heath and Company
Lexington, Massachusetts
Toronto

Library of Congress Cataloging in Publication Data

Merriam, Charles Wolcott, 1931–
 FORTRAN computer program.

 Includes bibliographical references and index. 1. Feedback control systems—Computer programs. 2. Mathematical optimization—Computer programs. 3. FORTRAN (computer program language) I. Title. TJ216.M4
629.8'3'0285425   77-14792
ISBN 0-669-01995-x

Copyright © 1978 by D.C. Heath and Company

All rights reserved. No part of this publication may be reproduced or transmitted in any form or by any means, electronic or mechanical, including photocopy, recording, or any information storage or retrieval system, without permission in writing from the publisher.

Published simultaneously in Canada

Printed in the United States of America

International Standard Book Number: 0-669-01995-x

Library of Congress Catalog Card Number: 77-14792

Contents

	List of Figures	vii
	List of Tables	ix
	Preface	xiii
Chapter 1	**General Programming Information**	1
	1.1 Introduction	1
	1.2 Format for Program Documentation	4
	1.3 Programming Conventions	5
Chapter 2	**Optimization with Functions**	9
	2.1 Steepest-Descent and Conjugate-Gradient Methods	9
	2.2 Newton-Raphson Method with Slack Variables	25
	2.3 Min-Max Method	42
Chapter 3	**Formulation of Performance Functionals**	69
	3.1 Excess-Pole-Specification Synthesis Method	69
	3.2 Evaluation of Desired Transfer-Function Matrix	72
Chapter 4	**Optimization with Functionals Defined on an Infinite Time Interval**	91
	4.1 Optimal Gain Control	91
	4.2 Optimal Linear Filters	114
	4.3 Simulation of Human Controller Models Based on Optimization	138
Chapter 5	**Optimization with Functionals Defined on a Finite Time Interval**	167
	5.1 Optimal Control Parameters with a Fixed Time Interval	167
	5.2 Optimal Control Functions with a Fixed Time Interval	176
	5.3 Optimal Control Functions with a Variable Time Interval	213

	5.4 Linearized Optimal Control with a Variable Time Interval	247
Chapter 6	**Utility Programs**	281
	6.1 Factoring Polynomials	281
	6.2 Solving Linear Algebraic Equations	323
	References	347
	Index	349
	About the Author	351

List of Figures

2-1	Computational flow chart for general iteration procedure	10
2-2	Computational flow chart for general trial procedure	12
6-1	Computational flow chart for subprogram FACTOR	285
6-2	Computational flow chart for subprogram FACT	293
6-3	Computational flow chart for subprogram INTL	300
6-4	Computational flow chart for subprogram ZEROS	305
6-5	Computational flow chart for subprogram SEARCH	311
6-6	Computational flow chart for subprogram PC	317

List of Tables

1-1	System-supplied programs and subprograms	3
1-2	User-supplied programs and subprograms	4
1-3	Maximum problem dimensions	5
2-1	Listing of MIN1	14
2-2a	Listing of user-supplied programs for Example 1	21
2-2b	Listing of user-supplied input data for Example 1	22
2-2c	Printed output from MIN1 for Example 1	22
2-3	Listing of MIN2	28
2-4a	Listing of user-supplied programs for Example 2A	36
2-4b	Listing of user-supplied input data for Example 2A	37
2-4c	Printed output from MIN2 for Example 2A	38
2-5	Listing of MIN3	45
2-6a	Listing of user-supplied programs for Example 2B	62
2-6b	Listing of user-supplied input data for Example 2B	63
2-6c	Printed output from MIN3 for Example 2B	64
3-1	Listing of SYN	73
3-2a	Listing of user-supplied input data for Example 3A	78
3-2b	Printed output from SYN for Example 3A	79
3-3	Listing of EVL	83
3-4a	Listing of user-supplied input data for Example 3B	88
3-4b	Printed output from EVL for Example 3B	88
4-1	Listing of FORM1	94
4-2a	Listing of user-supplied input data for Example 4A	99
4-2b	Printed output from FORM1 for Example 4A	100
4-3	Listing of QUAD	102
4-4a	Listing of user-supplied input data for Example 4B	106
4-4b	Printed output from QUAD for Example 4B	107
4-5	Listing of GAIN1	111
4-6a	Listing of user-supplied input data for Example 4C	114
4-6b	Printed output from GAIN1 for Example 4C	115
4-7	Listing of LOAD1	116
4-8a	Listing of user-supplied programs for Example 4D	121
4-8b	Listing of user-supplied input data for Example 4D	122
4-8c	Printed output from MIN1 and LOAD1 for Example 4D	123
4-9	Listing of FORM2	129
4-10a	Listing of user-supplied input data for Example 5A	134
4-10b	Printed output from FORM2 for Example 5A	135
4-11a	Listing of user-supplied input data for Example 5B	136
4-11b	Printed output from QUAD for Example 5B	137
4-12	Listing of GAIN2	139

4–13a	Listing of user-supplied input for Example 5C	142
4–13b	Printed output from GAIN2 for Example 5C	143
4–14	Listing of INT	146
4–15	Listing of DXM	150
4–16a	Listing of user-supplied programs for Example 6A	154
4–16b	Listing of User-supplied input data for Example 6A	156
4–16c	Printed output from INT and INTS for Example 6A	157
4–17	Listing of LOAD2	160
4–18a	Listing of user-supplied programs for Example 6B	163
4–18b	Listing of user-supplied input data for Example 6B	164
4–18c	Printed output from LOAD2 and MIN1 for Example 6B	165
5–1	Listing of LOAD3	170
5–2a	Listing of user-supplied programs for Example 7A	177
5–2b	Listing of user-supplied input data for Example 7A	178
5–2c	Printed output from MIN1 and LOAD3 for Example 7A	179
5–3	Listing of MIN4	186
5–4	Listing of FUN	193
5–5a	Listing of user-supplied programs for Example 7B	205
5–5b	Listing of user-supplied input data for Example 7B	206
5–5c	Printed output from MIN4 and FUN for Example 7B	207
5–6	Listing of MIN5	217
5–7a	Listing of user-supplied programs for Example 8A	237
5–7b	Listing of user-supplied input data for Example 8A	239
5–7c	Printed output from MIN5 for Example 8A	240
5–8	Listing of SIM	250
5–9a	Listing of user-supplied programs for Example 8B	268
5–9b	Listing of user-supplied input data for Example 8B	270
5–9c	Printed output from SIM for Example 8B	271
6–1	Listing of FACTOR	287
6–2	Listing of FACT	295
6–3	Listing of INTL	302
6–4	Listing of ZEROS	307
6–5	Listing of CONVEX	310
6–6	Listing of SEARCH	312
6–7	Listing of EVAL	314
6–8	Listing of MATCH	316
6–9	Listing of PC	318
6–10	Listing of PD	319
6–11	Listing of PM	320
6–12	Listing of TEST	322
6–13a	Listing of user-supplied input data for Example 9	324
6–13b	Printed output from TEST and subprograms for Example 9	325
6–14	Listing of LIN1	333

LIST OF TABLES

6-15	Listing of LIN1A	336
6-16	Listing of LIN2	340
6-17	Listing of INC	342
6-18	Listing of MI	343
6-19	Listing of MIA	345

Preface

This book is intended to provide the means for solving a wide range of computational problems which arise in the application of optimization techniques to the design of feedback control systems. FORTRAN IV computer programs are listed, and examples of their application are given. Short summaries of the underlying optimization theory and computational methods employed in the corresponding computer programs are also provided.

This book is organized in a pedagogic mode corresponding to a typical progression of concepts that might be followed in many courses on the application of optimization techniques to the design of feedback control systems. The organization is similar to my book *Automated Design of Control Systems* [1]. These books are supplementary and, together, form the basis for course work which blends theory with application.

The computer programs are written in the programming language FORTRAN IV, and they should run as is on all computer systems which support FORTRAN IV, with one possible exception. Program ZEROS is written with a subroutine call ERRSET (208, 256, -1, 1), which suppresses diagnostic procedures induced by exponent underflow. Subroutine ERRSET is a standard IBM 360/65 system routine which can be omitted or replaced by an appropriate system subroutine, such as TRAPS (0, 0, 1000) used for the WATFIV compiler. These programs, when compiled and linked with problem-oriented subroutines, also are efficient and can be used effectively in a production mode. A copy of these computer programs and the input corresponding to the examples given in this book is available on cards or tape from the author, upon request and upon payment for the costs of duplication and mailing. These computer programs have been written in a uniform style in order to facilitate their usage as the user progresses through the concepts presented in the book.

I am indebted to many people and organizations for their support in the preparation of this book. My long standing association with the General Electric Company, most recently with Mr. Edward Balbirnie of the Re-Entry and Environmental Systems Division in Philadelphia, has been the principal catalyst for the development and application of the production computer programs. The University of Rochester has also provided the academic atmosphere and opportunity to pursue a project with pedagogic objectives. Much of the computer costs of developing these computer programs were borne by the Center for Naval Analysis (CNA SUB N00014-68-A-0091). Finally, I am indebted to Patricia Saylor for her skillful and careful typing of the manuscript.

C.W. Merriam III

1

General Programming Information

1.1 Introduction

System-Supplied Programs and Subprograms

A system of thirty-seven FORTRAN IV computer programs for solving optimization problems arising in feedback control is described. These computer programs are implementations of many of the computational methods presented in the main body of *Automated Design of Control Systems* [1], and they can be conveniently used to implement additional computational methods also described in that book, as well as similar books such as *Applied Optimal Control* [2]. The computer programs supplied by the system are divided into four basic types.

    The first type is a main program that implements a computational method of major import to control optimization. For example, program QUAD implements a Newton method for solving the algebraic form of the matrix Riccati equation. There are eight such programs, namely, EVL, FORM1, FORM2, GAIN1, GAIN2, QUAD, SYN, and TEST. These programs do not have an argument list, require input data, and require no user-supplied subprograms. They should be compiled, fully linked to compiled subprograms which are called, and stored as an object module in the host computer system if they are to be used frequently by students or others.

    The second type is a subprogram that also implements a computational method of major import to control optimization. For example, program MIN1 implements both the steepest-descent and conjugate-gradient methods of successive approximation. There are seven such subprograms, namely, INT, MIN1, MIN2, MIN3, MIN4, MIN5, and SIM. These subprograms should be compiled and stored in unlinked form in the host computer system. They do not have an argument list, require input data, and require both an elementary user-supplied main program and a user-supplied subprogram that defines the particular problem to be solved. Once the appropriate main program and subprogram are made available by the user, linking and storing as an object module can then be accomplished by a relatively inexpensive LOADER, which is normally resident in the host computer library of systems programs.

    The third type is a subprogram that specializes a computational method of major import to control optimization to a broad class of design problems.

For example, subprogram LOAD1 can be used with either MIN1 or MIN3 to compute optimal gains for the design of linear control systems with deterministic signals. There are three such subprograms, namely, FUN, LOAD1, and LOAD3. These subprograms are just a few examples of a large number of possible general purpose, user-supplied subprograms that can be added to the system of programs described in this book. For example, a subprogram, which is similar to LOAD1, for specializing MIN1 or MIN3 to optimal gain computation for the design of linear control systems with stochastic signals could easily be supplied by a user. When needed, subprograms in this third category should be compiled and stored in linked object-module form with the calling program and other required programs for subsequent production usage with data. User-supplied subprograms for specializing general computational techniques are treated the same once they have been written. These subprograms have an argument list, generally require input data, require an elementary user-supplied main program, but generally do not require an additional user-supplied subprogram.

The fourth type is a subprogram supplied by the system as a utility. These subprograms are needed to implement one or more of the computational methods of major import to control optimization. For example, subprogram MI implements matrix inversion by the pivot method. There are nineteen such utility subprograms, namely, CONVEX, DXM, EVAL, FACT, FACTOR, INC, INTL, LIN1, LIN1A, LIN2, LOAD2, MATCH, MI, MIA, PC, PD, PM, SEARCH, and ZEROS. They require compilation and linking with main programs and other subprograms as needed.

A summary of system-supplied programs and subprograms is given in Table 1-1. Examples presented in this book, where these system programs and subprograms are used, are also listed in Table 1-1.

User-Supplied Programs and Subprograms

Both main programs and subprograms must be supplied by the user in order to use many of the computational methods implemented by the system of programs described. These user-supplied programs and subprograms (see Table 1-2) are named in a way that immediately identifies them with the appropriate system-supplied subprogram. For example, use of subprogram MIN1 requires a user-supplied main program MIN1C and a user-supplied subprogram MIN1S.

User-supplied main programs are elementary and serve merely to call the corresponding system-supplied subprogram. User-supplied subprograms may be fairly simple programs for defining a specific design problem. For example, subprogram MIN1S is used to compute function $f(\mathbf{c})$ and gradient $f_\mathbf{c}$ for execution with MIN1. However, user-supplied subprograms may be elaborate programs for defining a class of specific design problems. For example, subprograms

Table 1-1
System-Supplied Programs and Subprograms

Name	Main/ Sub	Argument List	Input Data	User-Supplied Sub	Example Usage
CONVEX	sub	yes	no	no	3B, 9
DXM	sub	yes	no	no	4B, 5B, 6A
EVAL	sub	yes	no	no	3B, 9
EVL	main	no	yes	no	3B
FACT	sub	yes	no	no	3B, 9
FACTOR	sub	yes	no	no	3B, 9
FORM1	main	no	yes	no	4A
FORM2	main	no	yes	no	5A
FUN	sub	yes	yes	no	7B
GAIN1	main	no	yes	no	4C
GAIN2	main	no	yes	no	5C
INC	sub	yes	no	no	4B, 5B, 6B
INT	sub	no	yes	yes	6A
INTL	sub	yes	no	no	3B, 9
LIN1	sub	yes	no	no	4D
LIN1A[a]	sub	yes	no	no	4D
LIN2	sub	yes	no	no	4B, 5B, 6B
LOAD1	sub	yes	yes	no	4D
LOAD2	sub	yes	no	no	6B
LOAD3	sub	yes	yes	yes	7A
MATCH	sub	yes	no	no	3B, 9
MI	sub	yes	no	no	2A, 2B, 3A, 4A, 4B, 4C, 4D, 5A, 5B, 5C, 6B, 8A, 8B
MIA[b]	sub	yes	no	no	2A, 2B, 3A, 4A, 4C, 5A, 5C, 8A, 8B
MIN1	sub	no	yes	yes	1, 4D, 6B, 7A
MIN2	sub	no	yes	yes	2A
MIN3	sub	no	yes	yes	2B
MIN4	sub	no	yes	yes	7B
MIN5	sub	no	yes	yes	8A
PC	sub	yes	no	no	3B, 9
PD	sub	yes	no	no	3B, 9
PM	sub	yes	no	no	9
QUAD	main	no	yes	no	4B, 5B
SEARCH	sub	yes	no	no	3B, 9
SIM	sub	no	yes	yes	8B
SYN	main	no	yes	no	3A
TEST	main	no	yes	no	9
ZEROS	sub	yes	no	no	3B, 9

[a] Can be substituted for LIN1 anywhere.

[b] Can be substituted for MI whenever the matrix to be inverted is symmetric positive-definite.

Table 1-2
User-Supplied Programs and Subprograms

Name	Main/Sub	Argument List	Example Usage
INTC	main	no	6A
INTS	sub	yes	6A
LOAD3S	sub	yes	7A
MIN1C	main	no	1, 4D, 6B, 7A
MIN1S	sub	yes	1, 4D, 6B, 7A
MIN2C	main	no	2A
MIN2S	sub	yes	2A
MIN3C	main	no	2B
MIN3S	sub	yes	2B
MIN4C	main	no	7B
MIN4S	sub	yes	7B
MIN5C	main	no	8A
MIN5S	sub	yes	8A
SIMC	main	no	8B
SIMS	sub	yes	8B

MIN4S and FUN, even though not user-supplied, illustrate subprograms that can be written to apply steepest-descent in one dimension to the solution of optimal control functions on a finite, fixed-time interval.

1.2 Format for Program Documentation

Each of the system-supplied programs and subprograms is documented in a uniform format as follows:

    1. Purpose. The general type of problem or function for which the program is intended is stated.

    2. Method. The computational method which is implemented by the program is stated. If appropriate, a general flow chart is also given for the method.

    3. Program description. (a) A dummy argument list is given for subprograms which have one. Corresponding definitions of program nomenclature are also given where appropriate. (b) An input data list is given for main programs and for subprograms which have one. Corresponding definitions of program nomenclature are also given where appropriate. (c) A list of output data is given for main programs and subprograms which have one. Output data can be routed to cards, tape, or disk as appropriate by the selection of job control statements. (d) A list of subprograms called is given, including user-supplied subprograms. Data produced by the user-supplied subprogram are also given where appropriate.

    4. User-supplied subprograms. User-supplied subprograms are described if any are called. The description includes a dummy argument list and the definition of corresponding nomenclature for the user's problem.

5. Example. Examples of each computational method of import to control optimization is given. There are nineteen such examples. Each example consists of the following: (a) problem statement; (b) user-supplied program and subprogram if needed; (c) list of input data cards; (d) printed data output.

The format of documentation described above should permit even a novice FORTRAN programmer to apply the system-supplied programs and subprograms to a large variety of control system design problems based on optimization techniques.

1.3 Programming Conventions

All programs in the system are subject to standard FORTRAN IV programming conventions [3]. Computer memory for these programs is allocated via dimension statements so that all of the examples and problems given in the main body of *Automated Design of Control Systems* [1] can be performed using this system. Specifically, maximum problem size is defined by Table 1-3 for FORTRAN programs supplied with this system. Maximum problem dimensions can be easily modified for the purpose of tailoring an efficient production program to the automated design of a broad class of control problems.

Additional programming conventions employed for this system of computer programs are summarized as follows:

1. Names of user-supplied subprograms end in the letter S. The remaining letters of these names are the same as the name of the system-supplied subprogram which calls the user-supplied subprogram. For example, subprogram MIN1S is called by subprogram MIN1.

2. Names of user-supplied main programs end in the letter C. The remaining letters of these names are the same as the name of the system-supplied

Table 1-3
Maximum Problem Dimensions

Type of Variable	Maximum Dimension
Parameters	21
Constraints	3
Controls	3
States	7
Responses	7
Pole specifications	7
Performance criteria	7
Observations	3
Additive noises	3
Multiplicative noises	3
Tabulated data points	401

subprogram which is called by the user-supplied calling program. For example, program MIN1C calls subprogram MIN1.

3. Variables with names beginning in D and DD are first and second derivatives, respectively. For example, DXV is the name of $\dot{x}$ and DFV is the name of f_c.

4. Variables ending in V are column vectors. For example, CV is the name of **c**.

5. Variables with names ending in M are matrices. For example, FM is the name of **F**.

6. Variables with names ending in A are three-dimensional arrays. For example, BA is the name of the array formed by placing $\mathbf{B}_1$ in the top horizontal plane, $\mathbf{B}_2$ in the next lower horizontal plane, etc. This three-dimensional array is denoted by $\mathbf{B}_1/\mathbf{B}_2/\ldots$ which identifies subscript i corresponding to $\mathbf{B}_i$ as the first of the three indices of an individual element. The last two indices of an array element correspond to standard matrix notation.

7. Variables with names ending in AA are four-dimensional arrays. For example, DBAA is the name of

$$(\mathbf{B}_1/\mathbf{B}_2/\ldots)_{c_1}/(\mathbf{B}_1\mathbf{B}_2/\ldots)_{c_2}/\ldots, \tag{1.1}$$

so that the first two indices of an element of this array are i and j corresponding to c_i and $\mathbf{B}_j$, respectively. The last two indices of an array element correspond to standard matrix notation.

8. Column vector input and output data are treated as a list of elements.

9. Matrix input and output data are treated as a list of elements using the standard FORTRAN convention of ordering elements by columns of the matrix.

10. Three-dimensional array input and output are treated as a sequence of matrices which is ordered according to the first index of the array.

11. Only the upper-triangular portions of symmetric matrices are used in input and output data.

12. Only the upper-triangular portions of symmetric matrices need be programmed in user-supplied subprograms. For example, matrix

$$f_{cc} = \begin{bmatrix} 1 & 0 & 1 \\ 0 & 1 & 0 \\ 1 & 0 & 1 \end{bmatrix} \tag{1.2}$$

can be defined by

$$f_{11} = f_{22} = f_{13} = f_{33} = 1 \text{ and } f_{12} = f_{23} = 0. \tag{1.3}$$

13. Quantities which are identically zero can be omitted in user-supplied subprograms. For instance, matrix elements

GENERAL PROGRAMMING INFORMATION

$$f_{12} = f_{23} = 0$$

need not be explicitly programmed for the previous example.

14. Character input data are keypunched using FORMAT (65A1).
15. Nonzero polynomials are expressed as

$$p(s) = \sum_{m=0}^{N_p - 1} p_{N_p - m} s^m, \qquad (1.4)$$

where $p_1 \neq 0$. The zero polynomial is defined similarly with $N_p = 1$ and $p_1 = 0$.

16. Integer variables are data type INTEGER*4 and are identified by the standard FORTRAN default convention for integer variables. They are keypunched using FORMAT (7I10) for input data.

17. Floating-point variables are data type REAL*8 and are keypunched using FORMAT (4D18.10) for input and output data.

18. Logical variables are keypunched using FORMAT (7G10) for input data.

System-supplied main programs and, in some cases, subprograms have card input and output data which are defined as follows: A comma delimits two adjacent fields on a card which are used for adjacent elements in an input data list. A slash delimits two adjacent input data lists which appear on separate sets of cards. Character input data lists used for comments are denoted as C1, C2, and C3 throughout. Input and output data are omitted when the data do not pertain to the problem being solved.

System-supplied subprograms are programmed as FORTRAN SUBROUTINES or FUNCTIONS and have a list of dummy arguments for input and output variables. A comma is used to delimit adjacent dummy arguments in this list.

Object-time dimensioning of user-supplied subprograms has been employed throughout this system of computer programs, and, hence, the user needs dimensioning information for programming user-supplied subprograms. Dimensions are specified in the subprogram description by the indices of the arrays being defined in the subsequent descriptions. For example, the specification

$$CV(NC) = c \qquad (1.5)$$

indicates that vector **c** has NC elements and is assigned the name CV.

User-supplied calling programs need not consist of more than three FORTRAN statements: the call to a subprogram, such as the statement CALL MIN1, the statement STOP, and the statement END.

2 Optimization with Functions

The computer programs presented in this chapter correspond to the computational methods given in chapter 3 of the book *Automated Design of Control Systems* [1]. These computer programs are implementations of computational methods used for minimizing functions with respect to a set of parameters. Equality and inequality constraints are also implemented.

The computational techniques presented in this chapter, as opposed to many other possibilities [4], are easily extended and applied to minimizing functionals with respect to control functions on variable time intervals. For example, the first program MIN1 presented in this chapter is applied to three additional types of design problems which are presented in subsequent chapters.

2.1 Steepest-Descent and Conjugate-Gradient Methods

The purpose of these computational methods is to minimize functions with respect to a set of unconstrained parameters. These methods require only computations of the objective function $f(\mathbf{c})$ and gradient vector $f_\mathbf{c}$.

Description of the Basic Methods

Steepest-descent and conjugate-gradient directions are examples of methods of computing vector $\mathbf{d}$ for the purpose of minimizing $f(\mathbf{c} + \alpha \mathbf{d})$ with respect to $\alpha > 0$. Each computation of a direction vector and corresponding minimization in one dimension is referred to as an iteration. A flow chart for a general iteration technique is given in Figure 2-1.

Let vector $\mathbf{f}$ be defined as the gradient

$$\mathbf{f} = f_\mathbf{c}. \tag{2.1}$$

Then direction vectors for this program can be expressed as

$$\mathbf{d}_i = -\mathbf{f}_i + [\operatorname{sgn}(i-\lfloor i/N \rfloor N)] \left[\frac{\mathbf{f}'_i \mathbf{f}_i}{\mathbf{f}'_{i-1} \mathbf{f}_{i-1}} \right] \mathbf{d}_{i-1}, \tag{2.2}$$

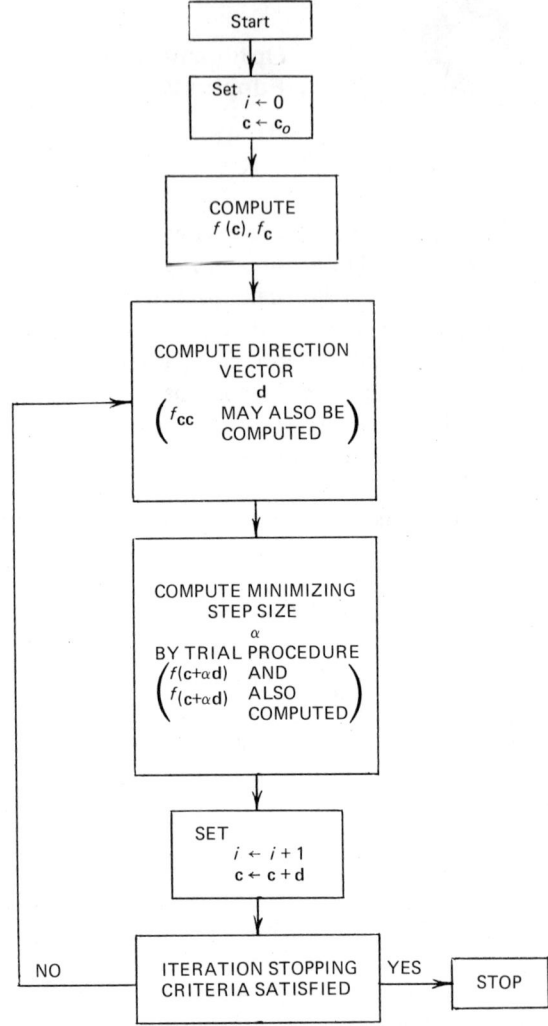

Reproduced from page 75 of *Automated Design of Control Systems* [1] by permission of Gordon and Breach, Science Publishers, Inc.

Figure 2-1. Computational Flow Chart for General Iteration Procedure

where $i = 0, 1, \ldots$ is the iteration number. Whenever index i is divisible by N, the corresponding direction vector is the steepest-descent direction. Therefore, condition $N = 1$ yields the steepest-descent method of successive approximations. Alternately, condition $N = \dim \mathbf{c}$ results in the method of conjugate directions which yields quadraticlike convergence, modulo N, when $\mathbf{c}_i$ is in a suitably small neighborhood of a relative minimum where hessian matrix f_{cc} is nonsingular.

Minimization of $f(c + \alpha d)$ with respect to $\alpha > 0$ is accomplished using the trial procedure depicted in Figure 2-2. This general trial procedure can be specialized easily by selecting initialization, extrapolation, and interpolation procedures. For example, quadratic extrapolation, based on an estimate f_{min} of the minimum value of function $f(c)$, results in the initialization procedure,

$$\alpha_o = 2 \frac{f_{min} - f(c)}{d' f_c}. \qquad (2.3)$$

Extrapolation and interpolation procedures can be expressed in the form

$$\alpha_+ = \beta + (\alpha - \beta)r, \qquad (2.4)$$

when ratio r is specified on the open interval (r_{min}, r_{max}) as follows: Quadratic extrapolation yields

$$r = \frac{b}{b - a}, \qquad (2.5)$$

where

$$s(\gamma) = \frac{\partial}{\partial \gamma} f(c + \gamma d), a = s(\alpha), \text{ and } b = s(\beta). \qquad (2.6)$$

Similarly, cubic interpolation yields

$$r = \frac{b + z + q}{b + a + 2z}, \qquad (2.7)$$

where

$$z = 3\left(\frac{b - a}{\alpha - \beta}\right) + b + a \quad \text{and} \quad q = [z - ab]^{\frac{1}{2}}. \qquad (2.8)$$

These methods of initialization, extrapolation, and interpolation require the computation of only function $f(c)$ and gradient f_c. Therefore, when the direction vector given in equation (2.2) is also used, successive approximations do not require computation of hessian matrix f_{cc}. Computation of this matrix becomes a very expensive proposition in many control system design problems of moderate dimensionality.

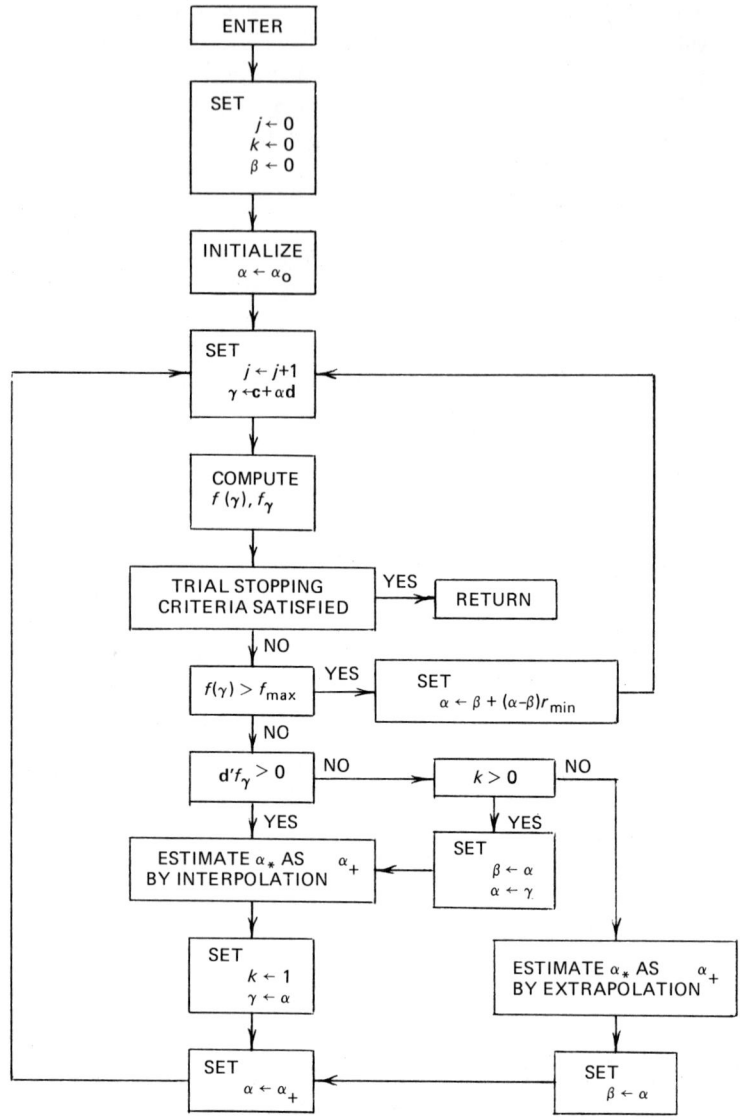

Reproduced from *Automated Design of Control Systems* [1] by permission of Gordon and Breach, Science Publishers, Inc., p. 77.

Figure 2-2. Computational Flow Chart for General Trial Procedure.

MIN1 and MIN1S

The trial and iterations procedures implemented by subprogram MIN1 are summarized as follows:

MIN1 function: Steepest-descent and conjugate-gradient methods of successive approximations for minimizing functions without constraints.

MIN1 dummy argument list: None.

MIN1 input data list: C1/C2/C3/NI,NT,NR,NC/RATIO,DFMIN,FMIN,FMAX/ ALMIN,ALMAX,RMIN,RMAX/CV, where

    NI = maximum number of iterations.
    NT = maximum number of trials per iteration.
    NR = number of iterations performed before restarting the conjugate-gradient method (1 gives steepest-descent).
    $1 \leq NC \leq 21$.
    RATIO = minimum value of $|s(\alpha)/s(0)|$ requiring additional trials.
    DFMIN = minimum value of $|f_c|$ requiring additional trials and iterations.
    FMIN = lower bound on $f(\mathbf{c})$.
    FMAX = upper bound on $f(\mathbf{c})$.
    ALMIN = minimum initial value of α.
    ALMAX = maximum initial value of α.
    RMIN = minimum fraction of $(\alpha - \beta)$ used in interval reduction (<1).
    RMAX = maximum fraction of $(\alpha - \beta)$ used in extrapolation (>1).
    CV(NC) = $\mathbf{c}$.

MIN1 output data list: CV, where CV(NC) = $\mathbf{c}$.

Subprograms called by MIN1: MIN1S for the computation of $F = f(\mathbf{c})$ and DFV(NC) = f_c.

Program MIN1 is listed in Table 2-1. User problems are specified for subprogram MIN1 by writing an appropriate subprogram MIN1S. Specifications for writing this subprogram are given as follows:

MIN1S function: Compute $f(\mathbf{c})$ and f_c for use with MIN1.
MIN1S dummy argument list: MC,CV,F,DFV,MODE, where
    CV(MC) = $\mathbf{c}$.
    F = $f(\mathbf{c})$.
    DFV(MC) = f_c.
    MODE = 1 for program initialization and 2 otherwise.

Table 2-1
Listing of MIN1

```
 1         SUBROUTINE MIN1
 2   C     REAL*8 CV(NC),DFV(NC),SCV(NC),DCV(NC)
 3         REAL*8 CV(21),DFV(21),SCV(21),DCV(21)
 4         REAL*8 A,ALMIN,ALMAX,B,C,DC,DF,DFA,DFB,DFMIN,X,
 5        1D1,D2,D3,F,FA,FB,FMIN,FMAX,RATIO,RMIN,RMAX,SDF,SVP
 6         REAL*8 DABS,DMIN1,DMAX1,DSQRT
 7         DATA MC/21/
 8         DATA F/0.D0/,DFV/21*0.D0/
 9         LOGICAL*1 COM(65)
10    1000 FORMAT(65A1)
11    1001 FORMAT(4X,65A1)
12    1002 FORMAT(4D18.10)
13    1003 FORMAT(4X,4D18.10)
14    1004 FORMAT(4I10)
15    1005 FORMAT('1')
16    1006 FORMAT(//,'               NI,NT,NR,NC')
17    1007 FORMAT('         RATIO,DFMIN,FMIN,FMAX')
18    1008 FORMAT('         ALMIN,ALMAX,RMIN,RMAX')
19    1009 FORMAT('         CV')
20    1010 FORMAT(//,'      ITERATION=',I2,'     TRIAL=',I2,'       F=',D18.10)
21    1011 FORMAT('   ALPHA=',D18.10,'   DF=',D18.10,'   |DFV|=',D18.10)
22    1012 FORMAT(//,'         RESET CV')
23    1013 FORMAT(///,'         DESIGN COMPLETE')
24    1014 FORMAT(///,'         ITERATION LIMIT')
```

(continued)

```
25  1015  FORMAT(//,'           TRIAL LIMIT')
26        WRITE (6,1005)
27        DO 1016 I=1,3
28        READ (5,1000) COM
29  1016  WRITE (6,1001) COM
30        READ (5,1004) NI,NT,NR,NC
31        READ (5,1002) RATIO,DFMIN,FMIN,FMAX
32        READ (5,1002) ALMIN,ALMAX,RMIN,RMAX
33        READ (5,1002) (CV(I),I=1,NC)
34        WRITE (6,1006)
35        WRITE (6,1004) NI,NT,NR,NC
36        WRITE (6,1007)
37        WRITE (6,1003) RATIO,DFMIN,FMIN,FMAX
38        WRITE (6,1008)
39        WRITE (6,1003) ALMIN,ALMAX,RMIN,RMAX
40        NIC=0
41        NTC=0
42        C=0.D0
43        NRC=NR
44        X=1.D0-1.D0/RMAX
45        CALL MIN1S(MC,CV,F,DFV,1)
46        CALL MIN1S(MC,CV,F,DFV,2)
47        DF=0.D0
48        DO 1017 I=1,NC
49        SCV(I)=CV(I)
50        DCV(I)=-DFV(I)
51  1017  DF=DF+DCV(I)*DFV(I)
```

Table 2-1 Continued

```
52         SVP=-DF
53         SDF=RATIO*DABS(DF)
54         D3=DSQRT(SVP)
55         WRITE (6,1010)  NIC,NTC,F
56         WRITE (6,1011)  C,DF,D3
57         WRITE (6,1009)
58         WRITE (6,1003)  (CV(I),I=1,NC)
59         IF (FMAX .GT. F) GO TO 1018
60         WRITE (6,1012)
61         RETURN
62   1018  IF (DFMIN .LT. D3) GO TO 1019
63         WRITE (6,1013)
64         WRITE (6,1005)
65         WRITE (7,1002)  (CV(I),I=1,NC)
66         RETURN
67   1019  DC=2.D0*(FMIN-F)/DF
68   1020  DC=DMAX1(ALMIN,DMIN1(DC,ALMAX))
69   1021  NTC=NTC+1
70         A=C
71         FA=F
72         DFA=DF
73         C=A+DC
74         DO 1022 I=1,NC
75   1022  CV(I)=SCV(I)+C*DCV(I)
76         CALL MIN1S(MC,CV,F,DFV,2)
```

OPTIMIZATION WITH FUNCTIONS

```
 77        DF=0.D0
 78        D1=0.D0
 79        DO 1023 I=1,NC
 80        DF=DF+DCV(I)*DFV(I)
 81   1023 D1=D1+DFV(I)*DFV(I)
 82        D2=DSQRT(D1)
 83        IF (SDF .GE. DABS(DF)) GO TO 1039
 84        WRITE (6,1010) NIC, NTC,F
 85        WRITE (6,1011) C,DF,D2
 86        WRITE (6,1009)
 87        WRITE (6,1003) (CV(I),I=1,NC)
 88        IF (FMAX .GT. F) GO TO 1026
 89   1024 IF (NT .GT. NTC) GO TO 1025
 90        GO TO 1030
 91   1025 DC=RMIN*DC
 92        C=A
 93        F=FA
 94        DF=DFA
 95        GO TO 1021
 96   1026 IF (DFMIN .LT. D2) GO TO 1027
 97        WRITE (6,1013)
 98        GO TO 1042
 99   1027 IF (DF) 1028,1028,1031
100   1028 IF (NT .LE. NTC) GO TO 1030
101        IF (X*DFA .LE. DF) GO TO 1029
102        DC=DC*RMAX
103        GO TO 1021
104   1029 DC=DC*DFA/(DFA-DF)
```

(continued)

Table 2-1 Continued

```
105            GO TO 1021
106     1030   WRITE (6,1015)
107            GO TO 1042
108     1031   B=C
109            FB=F
110            DFB=DF
111     1032   D1=DFA+DFB+3.D0*(FA-FB)/DC
112            D2=2.D0*(DFA+D1)
113            D3=DFA+DFB+2.D0*D1
114            IF (1.D-5*DABS(D2) .LE. DABS(D3)) GO TO 1033
115            C=B-(DFA+2.D0*D1)*DC/D2
116            GO TO 1034
117     1033   D2=DSQRT(D1**2-DFA*DFB)
118            C=B-(DFB+D1-D2)*DC/D3
119     1034   DO 1Q35 I=1,NC
120     1035   CV(I)=SCV(I)+C*DCV(I)
121            CALL MIN1S(MC,CV,F,DFV,2)
122            DF=0.D0
123            D1=0.D0
124            DO 1036 I=1,NC
125            DF=DF+DCV(I)*DFV(I)
126     1036   D1=D1+DFV(I)*DFV(I)
127            D2=DSQRT(D1)
128            IF (SDF .GE. DABS(DF)) GO TO 1039
129            NTC=NTC+1
```

(continued)

```
130            WRITE (6,1010) NIC,NTC,F
131            WRITE (6,1011) C,DF, D2
132            WRITE (6,1009)
133            WRITE (6,1003)   (CV(I),I=1,NC)
134            IF (FMAX .LE. F) GO TO 1024
135            IF (NT .LE. NTC) GO TO 1030
136            IF (DF)    1038,1037,1037
137     1037   DC=C-A
138            GO TO 1031
139     1038   A=C
140            FA=F
141            DFA=DF
142            DC=B-C
143            GO TO 1032
144     1039   NIC=NIC+1
145            DC=C
146            C=0.D0
147            NTC=0
148            IF (NRC .GT. NIC) GO TO 1040
149            SVP=1.D+37
150            NRC=NRC+NR
151     1040   B=D1/SVP
152            SVP=D1
153            DF=0.D0
154            DO 1041 I=1,NC
155            SCV(I)=CV(I)
156            DCV(I)=B*DCV(I)-DFV(I)
```

Table 2-1 Continued

```
157      1041 DF=DF+DCV(I)*DFV(I)
158           SDF=RATIO*DABS(DF)
159           WRITE (6,1010) NIC,NTC,F
160           WRITE (6,1011) C,DF,D2
161           WRITE (6,1009)
162           WRITE (6,1003) (CV(I),I=1,NC)
163           IF (NI .GT. NIC) GO TO 1020
164           WRITE (6,1014)
165      1042 WRITE (6,1005)
166           WRITE (7,1002) (SCV(I),I=1,NC)
167           RETURN
168           END
```

Example 1

An elementary example of the use of subprogram MIN1 is the minimization of

$$f(c) = 1 - c'f + \frac{1}{2}c'Fc$$

by the conjugate-gradient method, where

$$f = \begin{bmatrix} 0 \\ 1 \\ 2 \end{bmatrix} \text{ and } F = \begin{bmatrix} 1 & 1 & 1 \\ 1 & 2 & 2 \\ 1 & 2 & 3 \end{bmatrix}.$$

Calling program MIN1C and the subprogram MIN1S corresponding to the above example must be supplied by the user; they are listed in Table 2-2a. A listing of program input cards for this example is given in Table 2-2b, and condition $N =$

Table 2-2a
Listing of User-supplied Programs for Example 1

```
1     C     MINC   EXAMPLE 1
2           CALL MIN1
3           STOP
4           END
5           SUBROUTINE MIN1S(MC,CV,F,DFV,MODE)
6           REAL*8 CV(MC),DFV(MC),F
7           REAL*8 FM(3,3),FV(3),DV(3),G,D
8           DATA FV/0.0D0,1.0D0,2.0D0/
9           DATA FM/4*1.0D0,2*2.0D0,1.0D0,2.0D0,3.0D0/
10          GO TO (1003,1000),MODE
11    1000  F=0.D0
12          G=0.D0
13          DO 1002 I=1,3
14          D=0.D0
15          DO 1001 J=1,3
16    1001  D=D+FM(I,J)*CV(J)
17          F=F+FV(I)*CV(I)
18          DFV(I)=D-FV(I)
19    1002  G=G+CV(I)*D
20          F=1.D0-F+5.D-1*G
21    1003  RETURN
22          END
```

Table 2-b
Listing of User-supplied Input Data for Example 1

```
1  MIN1(MIN1S)    EXAMPLE 1
2  MERRIAM
3  UNIVERSITY OF ROCHESTER
4       3            4              3              3
5     0.01        .0000001         0.             5.
6     0.1           10.            0.5           10.
7     1.            0.             0.
```

Table 2-2c
Printed Output from MIN1 for Example 1

```
MIN1(MIN1S)    EXAMPLE 1
MERRIAM
UNIVERSITY OF ROCHESTER

NI,NT,NR,NC
   3        4              3              3
RATIO,DFMIN,FMIN,FMAX
 0.1000000000D-01  0.1000000000D-06  0.0000000000D 00  0.5000000000D 01
ALMIN,ALMAX,RMIN,RMAX
 0.1000000000D 00  0.1000000000D 02  0.5000000000D 00  0.1000000000D 02
```

OPTIMIZATION WITH FUNCTIONS

```
ITERATION= 0    TRIAL= 0    F=  0.150000000D 01
ALPHA=  0.000000000D 00    DF= -0.200000000D 01    |DFV|=  0.141421356 2D 01
CV
    0.100000000D 01   0.000000000D 00   0.000000000D 00


ITERATION= 0    TRIAL= 1    F=  0.750000000D 00
ALPHA=  0.150000000D 01    DF=  0.100000000D 01    |DFV|=  0.269258240 4D 01
CV
   -0.500000000D 00   0.000000000D 00   0.150000000D 01


ITERATION= 1    TRIAL= 0    F=  0.500000000D 00
ALPHA=  0.000000000D 00    DF= -0.300000000D 01    |DFV|=  0.173205080 8D 01
CV
    0.000000000D 00   0.000000000D 00   0.100000000D 01


ITERATION= 1    TRIAL= 1    F=  0.225000000D 01
ALPHA=  0.100000000D 01    DF=  0.650000000D 01    |DFV|=  0.377491721 8D 01
CV
   -0.250000000D 01  -0.100000000D 01   0.150000000D 01


ITERATION= 2    TRIAL= 0    F=  0.263157894 7D-01
ALPHA=  0.000000000D 00    DF= -0.166204986 1D-01   |DFV|=  0.128920512 8D 00
CV
   -0.789473684 2D 00 +0.315789473 7D 00   0.115789473 7D 01
```

(continued)

23

Table 2-2c Continued

```
ITERATION= 2   TRIAL= 1    F=  0.2132891294D-01
ALPHA=  0.3157894737D 00   DF= -0.1496305277D-01   |DFV|=  0.1160641736D 00
CV
  -0.8104679983D 00  -0.2842980026D 00   0.1142149001D 01

ITERATION= 2   TRIAL= 2    F=  0.2473649075D-03
ALPHA=  0.3473684211D 01   DF=  0.1611405683D-02   |DFV|=  0.1249921869D-01
CV
  -0.1020411139D 01   0.3061670797D-01   0.9846916460D 00

ITERATION= 3   TRIAL= 0    F=  0.0000000000D 00
ALPHA=  0.0000000000D 00   DF= -0.6900375879D-28   |DFV|=  0.8306850112D-14
CV
  -0.1000000000D 01  -0.1637578961D-14   0.1000000000D 01

ITERATION LIMIT
```

dimc (i.e., NR = NC) has been established by these input data. Finally, program printed output is listed in Table 2-2c for this example. These data illustrate three-step convergence of the conjugate-gradient method for a quadratic function of three variables. The reader can also use this example to compare the performance of steepest-descent and conjugate-gradient methods when applied to quadratic functions.

2.2 Newton-Raphson Method with Slack Variables

The purpose of this computational method is to minimize functions with respect to a set of parameters which may be constrained by algebraic equalities or inequalities. This method requires the computation of objective function $f(\mathbf{c})$, gradient vector $f_\mathbf{c}$, hessian matrix $f_\mathbf{cc}$, constraint vector $\mathbf{g}(\mathbf{c})$, and Jacobian matrix $\mathbf{g_c}$.

Description of the Basic Method

Inequality constraints of the form

$$\mathbf{g(c)} \leqslant \mathbf{0} \qquad (2.9)$$

can be converted to equivalent and equality constraints

$$\mathbf{g(c)} + \sigma(\mathbf{s}) = \mathbf{0}, \qquad (2.10)$$

where, for example, the vector

$$\sigma(\mathbf{s}) = \begin{bmatrix} s_1^2/2 \\ s_2^2/2 \\ \vdots \\ \vdots \end{bmatrix} \qquad (2.11)$$

is introduced so that $\sigma(\mathbf{s}) \geqslant \mathbf{0}$. Slack vector $\mathbf{s}$ then can be regarded as a continuation of the parameter vector $\mathbf{c}$.

An equivalent objective function, the Lagrangian, can now be introduced for the equality constraints given in equation (2.10) as

$$L = f(\mathbf{c}) + \lambda'[\mathbf{g(c)} + \sigma(\mathbf{s})], \qquad (2.12)$$

where λ is a vector of Lagrange multipliers. The Newton-Raphson method for this equivalent objective function is given by

$$\begin{bmatrix} c \\ s \\ \lambda \end{bmatrix}_{i+1} = \begin{bmatrix} c \\ s \\ \lambda \end{bmatrix}_i - \begin{bmatrix} L_{cc} & 0 & L_{c\lambda} \\ 0 & L_{ss} & L_{s\lambda} \\ L_{\lambda c} & L_{\lambda s} & 0 \end{bmatrix}_i^{-1} \begin{bmatrix} L_c \\ L_s \\ L_\lambda \end{bmatrix}_i \quad (2.13)$$

for $i = 0, 1, \ldots$. This method of successive substitutions requires the computation of vectors

$$L_c = f_c + (\lambda' g)_c, \quad L_s = (\text{diag } s)\lambda,$$
$$L_\lambda = g + \sigma \quad (2.14)$$

and matrices

$$L_{cc} = f_{cc} + (\lambda' g)_{cc}, \quad L_{ss} = \text{diag } \lambda,$$
$$L_{\lambda c} = g_c, \quad L_{\lambda s} = \text{diag } s. \quad (2.15)$$

Equality constraints of the form $g(c) = 0$ can be interspersed with inequality constraints by simply eliminating the corresponding slack variables that appear in equation (2.13).

The Newton-Raphson method of successive substitutions does not require extensive logic for iterations and trials; thus, no flow chart is given. However, matrix L_{cc} of second partial derivatives does have to be computed, and this computation is very expensive in many control system design problems of moderate dimensionality. Moreover, the Newton-Raphson method, particularly with inequality constraints, may converge to any one of a number of limit points which do not correspond to the problem intended. Therefore, the subsequent computer program implementation of the Newton-Raphson method with inequality constraints includes a facility to omit inequality constraints, which are ineffective, via program control based on input data.

MIN2 and MIN2S

The successive substitution procedure outlined previously in this section is implemented by subprogram MIN2 which is summarized as follows:

MIN2 function: Newton-Raphson method of successive substitutions for minimizing functions with equality and inequality constraints.

OPTIMIZATION WITH FUNCTIONS

MIN2 dummy argument list: None.

MIN2 input data list: C1/C2/C3/NI,NC,NEC,NIC/DFMIN/CV/LAMV/SV/ IV/MAXV, where
   NI = maximum number of iterations.
   $1 \leq NC \leq 21, 0 \leq NEC \leq 3-NIC, 0 \leq NIC \leq 3-NEC$.
   DFMIN = minimum value of $|L_c|$ requiring additional iterations.
   CV(NC) = c.
   LAMV(NEC+NIC) = λ.
   SV(NIC) = s.
   IV(NIC) = logical vector of effective (*T*) and ineffective (*F*) inequality constraints.
   MAXV(NC) = vector of maximum permissible magnitudes of increments in the elements of c.

MIN2 output data list: CV/LAMV/SV, where
   CV(NC) = c.
   LAMV(NEC+NIC) = λ.
   SV(NIC) = s.

Subprograms called by MIN2: MI and MIN2S for the computation of $F = f(\mathbf{c})$, DFV(NC) = $f_\mathbf{c}$, DDFM(NC,NC) = $f_{\mathbf{cc}}$, GV(NEC+NIC) = $\mathbf{g}(\mathbf{c})$, DGM(NEC+NIC,NC) = $\mathbf{g_c}$, and DDGA (NEC+NIC,NC,NC) = $(g_1)_{\mathbf{cc}}/(g_2)_{\mathbf{cc}}/\ldots$

Program MIN2 is listed in Table 2-3. User problems are specified for subprogram MIN2 by writing an appropriate subprogram MIN2S. Specifications for writing this subprogram are given as follows:

MIN2S function: Compute $f(\mathbf{c}), f_\mathbf{c}, f_{\mathbf{cc}}, \mathbf{g}(\mathbf{c}), \mathbf{g_c}$, and $(g_1)_{\mathbf{cc}}/(g_2)_{\mathbf{cc}}/\ldots$ for use with MIN2.

MIN2S dummy argument list: MC,MG,CV,F,DFV,DFM,GV,DDGM,DDGA, MODE, where
   CV(MC) = c.
   F = $f(\mathbf{c})$.
   DFV(MC) = $f_\mathbf{c}$.
   DDFM(MC,MC) = $f_{\mathbf{cc}}$.
   GV(MG) = $\mathbf{g}(\mathbf{c})$.
   DGM(MG,MC) = $\mathbf{g_c}$.
   DDGA(MG,MC,MC) = $(g_1)_{\mathbf{cc}}/(g_2)_{\mathbf{cc}}/\ldots$
   MODE = 1 for program initialization and 2 otherwise.

Table 2-3
Listing of MIN2

```
1           SUBROUTINE MIN2
2     C     REAL*8 DDGA(N1,NC,NC),DDFM(NC,NC),DGM(N1,NC),LM(N4,N4),CV(NC),
3     C    1LAMV(N1),SV(NIC),MDCV(NC),GV(N1),DFV(NC),LV(N4),HV(N4)
4     C     LOGICAL*1 EICV(NIC),IV(N4)
5           REAL*8 DDGA(3,21,21),DDFM(21,21),DGM(3,21),LM(30,30),CV(21),
6          1LAMV(3),SV(3),MDCV(21),GV(3),DFV(21),LV(30),HV(30)
7           LOGICAL*1 EICV(3),IV(30)
8           REAL*8 ALPHA,DFMIN,F,D,L
9           REAL*8 DABS,DSQRT
10          DATA MC/21/,M1/3/,MT/30/
11          DATA F/0.D0/,DFV/21*0.D0/,DDFM/441*0.D0/,GV/3*0.D0/,
12         1DGM/63*0.D0/,DDGA/1323*0.D0/
13          LOGICAL*1 COM(65)
14    1000  FORMAT(65A1)
15    1001  FORMAT(4X,65A1)
16    1002  FORMAT(4D18.10)
17    1003  FORMAT(4I10)
18    1004  FORMAT(7G10)
19    1005  FORMAT(4X,4D18.10)
20    1006  FORMAT('1')
21    1007  FORMAT(//,'        NI,NC,NEC,NIC')
22    1008  FORMAT('       DFMIN')
23    1009  FORMAT('       CV')
24    1010  FORMAT('       LAMV')
```

```
25   1011 FORMAT('          SV')
26   1012 FORMAT('          IV')
27   1013 FORMAT('          MAXV')
28   1014 FORMAT(//,'            ITERATION=',I2,'     F=',D18.10,'   L=',D18.10)
29   1015 FORMAT('            ALPHA=',D18.10,'   |LV|=',D18.10)
30   1016 FORMAT('          GV')
31   1017 FORMAT(//,'DESIGN COMPLETE')
32   1018 FORMAT(//,'ITERATION LIMIT')
33        WRITE (6,1006)
34        DO 1019 I=1,3
35        READ (5,1000) COM
36   1019 WRITE (6,1001) COM
37        READ (5,1003) NI,NC,NEC,NIC
38        READ (5,1002) DFMIN
39        READ (5,1002) (CV(I),I=1,NC)
40        WRITE (6,1007)
41        WRITE (6,1003) NI,NC,NEC,NIC
42        WRITE (6,1008)
43        WRITE (6,1005) DFMIN
44        N1=NEC+NIC
45        N2=NC+NEC
46        N3=N2+NIC
47        N4=N3+NIC
48        IF (N1) 1022,1022,1020
49   1020 READ (5,1002) (LAMV(I),I=1,N1)
50        IF (NIC) 1022,1022,1021
51   1021 READ (5,1002) (SV(I),I=1,NIC)
52        READ (5,1004) (EICV(I),I=1,NIC)
```

(continued)

Table 2-3 Continued

```
53            WRITE (6,1012)
54            WRITE (6,1004)   (EICV(I),I=1,NIC)
55     1022   READ (5,1002)    (MDCV(I),I=1,NC)
56            WRITE (6,1013)
57            WRITE (6,1005)   (MDCV(I),I=1,NC)
58            DO 1023 I=1,N4
59     1023   LV(I)=0.D0
60            N=N2
61            DO 1024 I=1,N2
62     1024   IV(I)=.TRUE.
63            IF (NIC) 1027,1027,1025
64     1025   DO 1026 I=1,NIC
65            IV(I+N2)=EICV(I)
66            IV(I+N3)=EICV(I)
67            IF (.NOT. EICV(I)) GO TO 1026
68            N=N+2
69     1026   CONTINUE
70     1027   NCC=0
71            CALL MIN2S(MC,M1,CV,F,DFV,DDFM,GV,DGM,DDGA,1)
72            ALPHA=0.D0
73            DO 1029 I=1,NC
74     1029   CV(I)=CV(I)+ALPHA*LV(I)
75            IF (N1) 1034,1034,1030
76     1030   DO 1031 I=1,N1
77     1031   LAMV(I)=LAMV(I)+ALPHA*LV(I+NC)
```

(continued)

```
 78          IF (NIC) 1034,1034,1032
 79  1032    DO 1033 I=1,NIC
 80  1033    SV(I)=SV(I)+ALPHA*LV(I+N3)
 81  1034    CALL MIN2S(MC,M1,CV,F,DFV,DDFM,GV,DGM,DDGA,2)
 82          L=F
 83          DO 1035 I=1,NC
 84  1035    LV(I)=DFV(I)
 85          IF (N1) 1046,1046,1036
 86  1036    DO 1037 I=1,N1
 87  1037    HV(I)=GV(I)
 88          IF (NIC) 1040,1040,1038
 89  1038    DO 1039 I=1,NIC
 90  1039    HV(I+NEC)=HV(I+NEC)+5.D-1*SV(I)**2
 91  1040    DO 1042 I=1,NC
 92          D=LV(I)
 93          DO 1041 J=1,N1
 94  1041    D=D+LAMV(J)*DGM(J,I)
 95  1042    LV(I)=D
 96          DO 1043 I=1,N1
 97          L=L+LAMV(I)*HV(I)
 98  1043    LV(I+NC)=HV(I)
 99          IF (NIC) 1046,1046,1044
100  1044    DO 1045 I=1,NIC
101  1045    LV(I+N3)=SV(I)*LAMV(I+NEC)
102  1046    D=Q.DO
103          DO 1047 I=1,N4
104  1047    D=D+LV(I)*LV(I)
```

Table 2-3 Continued

```
105            D=DSQRT(D)
106            WRITE (6,1014) NCC,F,L
107            WRITE (6,1015) ALPHA,D
108            WRITE (6,1009)
109            WRITE (6,1005) (CV(I),I=1,NC)
110            IF (N1) 1050,1050,1048
111       1048 WRITE (6,1016)
112            WRITE (6,1005) (GV(I),I=1,N1)
113            WRITE (6,1010)
114            WRITE (6,1005) (LAMV(I),I=1,N1)
115            IF (NIC) 1050,1050,1049
116       1049 WRITE (6,1011)
117            WRITE (6,1005) (SV(I),I=1,NIC)
118       1050 IF (DFMIN .LT. D) GO TO 1051
119            WRITE (6,1017)
120            GO TO 1052
121       1051 NCC=NCC+1
122            IF (NCC .LE. NI) GO TO 1056
123            WRITE (6,1018)
124       1052 WRITE (6,1006)
125            WRITE (7,1002) (CV(I),I=1,NC)
126            IF (N1) 1055,1055,1053
127       1053 WRITE (7,1002) (LAMV(I),I=1,N1)
128            IF (NIC) 1055,1055,1054
129       1054 WRITE (7,1002) (SV(I),I=1,NIC)
```

```
130  1055  RETURN
131  1056        DO 1059 I=1,NC
132              DO 1059 J=I,NC
133              D=DDFM(I,J)
134              IF (N1) 1059,1059,1057
135  1057        DO 1058 K=1,N1
136  1058        D=D+LAMV(K)*DDGA(K,I,J)
137  1059        LM(I,J)=D
138              IF (N1) 1068,1068,1060
139  1060        DO 1062 J=1,N1
140              K=J+NC
141              DO 1061 I=1,NC
142  1061        LM(I,K)=DGM(J,I)
143              DO 1062 I=1,J
144  1062        LM(I+NC,K)=0.D0
145              IF (NIC) 1068,1068,1063
146  1063        DO 1067 J=1,NIC
147              K=J+N3
148              DO 1064 I=1,N2
149  1064        LM(I,K)=0.D0
150              DO 1065 I=1,NIC
151  1065        LM(I+N2,K)=0.D0
152              LM(J+N2,K)=SV(J)
153              DO 1066 I=1,J
154  1066        LM(I+N3,K)=0.D0
155  1067        LM(K,K)=LAMV(J+NEC)
156  1068        K=0
```

(continued)

Table 2-3 Continued

```
157            DO 1070 J=1,N4
158            IF (.NOT. IV(J)) GO TO 1070
159            K=K+1
160            LV(K)=LV(J)
161            M=0
162            DO 1069 I=1,J
163            IF (.NOT. IV(I)) GO TO 1069
164            M=M+1
165            LM(M,K)=LM(I,J)
166            LM(K,M)=LM(I,J)
167   1069  CONTINUE
168   1070  CONTINUE
169            CALL MI(MT,N,LM)
170            DO 1072 I=1,N
171            D=0.D0
172            DO 1071 J=1,N
173   1071  D=D-LM(I,J)*LV(J)
174   1072  HV(I)=D
175            K=0
176            DO 1074 I=1,N4
177            IF (IV(I)) GO TO 1073
178            LV(I)=0.D0
179            GO TO 1074
180   1073  K=K+1
181            LV(I)=HV(K)
```

```
182   1074 CONTINUE
183        ALPHA=1.D0
184        DO 1075 I=1,NC
185        D =DABS(LV(I))
186        IF (ALPHA*D .LE. MDCV(I)) GO TO 1075
187        ALPHA=MDCV(I)/D
188   1075 CONTINUE
189        GO TO 1028
190        END
```

Table 2-4a
Listing of User-supplied Programs for Example 2A

```
1      C      MIN2C    EXAMPLE  2A
2             CALL MIN2
3             STOP
4             END
5             SUBROUTINE MIN2S(MC,M1,CV,F,DFV,DDFM,GV,DGM,DDGA,MODE)
6             REAL*8 DDGA(M1,MC,MC),DGM(M1,MC),DDFM(MC,MC),CV(MC),
7            1DFV(MC),GV(M1),F
8             REAL*8 FV(2),D
9             DATA FV/1.D0,2.D0/
10            GO TO (1000,1001),MODE
11      1000  DDFM(1,1)=2.D0
12            DDFM(2,2)=4.D0
13            DDFM(3,3)=6.D0
14            DGM(1,1)=1.D0
15            DGM(2,1)=1.D0
16            DGM(1,2)=1.D0
17            DGM(2,2)=2.D0
18            DGM(1,3)=1.D0
19            DGM(2,3)=1.D0
20            RETURN
21      1001  F=CV(1)**2+2.D0*CV(2)**2+3.D0*CV(3)**2
22            DFV(1)=2.D0*CV(1)
23            DFV(2)=4.D0*CV(2)
24            DFV(3)=6.D0*CV(3)
25            DO 1003 I=1,2
26            D=FV(I)
27            DO 1002 J=1,3
28      1002  D=D+DGM(I,J)*CV(J)
29      1003  GV(I)=D
30            RETURN
```

Example 2A

An elementary example of the use of subprogram MIN2 is the minimization of

$$f(c) = c'Fc \text{ subject to } g + Gc \leq 0,$$

where

$$F = \begin{bmatrix} 1 & 0 & 0 \\ 0 & 2 & 0 \\ 0 & 0 & 3 \end{bmatrix}, g = \begin{bmatrix} 1 \\ 2 \end{bmatrix}, G = \begin{bmatrix} 1 & 1 & 1 \\ 1 & 2 & 1 \end{bmatrix}.$$

Calling program MIN2C and the subprogram MIN2S corresponding to the above example must be supplied by the user; they are listed in Table 2-4a. A listing of program input cards for this example is given in Table 2-4b, and both constraints have been made effective via input data for vector IV. Finally, program printed output is listed in Table 2-4c for this example. These data illustrate quadratic convergence of the Newton-Raphson technique using slack variables for inequality constraints. Note that the first constraint is determined to be indeed ineffective and hence could have been eliminated, thereby simplifying the selection of c_0, s_0, and λ_0. This example can be used by the reader to discover how difficult the initialization problem is for this method and what kind of limit points result from other initializations.

Table 2-4b
Listing of User-supplied Input Data for Example 2A

```
 1    MIN2(MI,MIN2S)     EXAMPLE 2A
 2    MERRIAM
 3    UNIVERSITY OF ROCHESTER
 4              6              3              0              2
 5        0.
 6        3.                  -3.                            0.
 7        1.                   1.
 8        1.                   1.
 9            T              T
10       10.                 10.                           10.
```

Table 2-4c
Printed Output from MIN2 for Example 2A

```
MIN2(MI,MIN2S)    EXAMPLE 2A
MERRIAM
UNIVERSITY OF ROCHESTER


NI,NC,NEC,NIC
  6    3         0              2
DFMIN
  0.000000000D 00
IV         T         T
MAXV
  0.100000000D 02  0.100000000D 02  0.100000000D 02  0.100000000D 02

ITERATION= 0      F= 0.270000000D 02      L= 0.280000000D 02
ALPHA=  0.000000000D 00     |LV|= 0.123895116 9D 02
CV
  0.300000000D 01 -0.300000000D 01  0.000000000D 00
GV
  0.100000000D 01 -0.100000000D 01
LAMV
  0.100000000D 01  0.100000000D 01
SV
  0.100000000D 01  0.100000000D 01
```

OPTIMIZATION WITH FUNCTIONS

```
ITERATION= 1    F=  0.8598765432D 00    L=  0.1138746914D 01
ALPHA=  0.1000000000D 01    |LV|=  0.4190768200D 00
CV
   -0.5500000000D 00   -0.4777777778D 00   -0.1833333333D 00
GV
   -0.2111111111D 00    0.3111111111D 00
LAMV
    0.2888888889D 00    0.8111111111D 00
SV
    0.7111111111D 00    0.1888888889D 00

ITERATION= 2    F=  0.1157377344D 01    L=  0.1206297193D 01
ALPHA=  0.1000000000D 01    |LV|=  0.1379703412D 00
CV
   -0.5793076916D 00   -0.5957827380D 00   -0.1931025639D 00
GV
   -0.3681929935D 00    0.3602426847D-01
LAMV
   -0.6590018564D-01    0.1224515569D 01
SV
    0.8733269527D 00   -0.9627227098D-01

ITERATION= 3    F=  0.1194700331D 01    L=  0.1199993174D 01
ALPHA=  0.1000000000D 01    |LV|=  0.5296315820D-02
CV
   -0.5988769533D 00   -0.5985380205D 00   -0.1996256511D 00
```

(continued)

Table 2-4c Continued

```
GV
    -0.3970406248D 00   0.4421354716D-02
LAMV
     0.1355731278D-02   0.1196398175D 01
SV
     0.8912934676D 00  -0.2210609161D-02

ITERATION= 4    F=  0.1199997050D 01   L=  0.1200000000D 01
ALPHA=  0.1000000000D 01   |LV|=  0.1059732355D-04
CV
    -0.5999985469D 00  -0.5999997397D 00  -0.1999995156D 00
GV
    -0.3999978022D 00   0.2458116018D-05
LAMV
    -0.4771282052D-05   0.1200001865D 01
SV
     0.8944302341D 00   0.6658610608D-05

ITERATION= 5    F=  0.1200000000D 01   L=  0.1200000000D 01
ALPHA=  0.1000000000D 01   |LV|=  0.2962986368D-10
CV
    -0.6000000000D 00  -0.6000000000D 00  -0.2000000000D 00
```

OPTIMIZATION WITH FUNCTIONS

```
GV
      -0.400000000D 00    0.221687390 7D-10
LAMV
      -0.162333412 6D-10  0.120000000 0D 01
SV
       0.894427191 0D 00  0.103487397 2D-10

ITERATION=  6    F=  0.120000000 0D 01    L=  0.120000000 0D 01
ALPHA=  0.100000000 0D 01   |LV|=  0.684427777D-15
CV
      -0.600000000 0D 00 -0.600000000 0D 00 -0.200000000 0D 00
GV
      -0.400000000 0D 00 -0.555111512 3D-16
LAMV
       0.253870131 0D-21  0.120000000 0D 01
SV
       0.894427191 0D 00 -0.167126430 5D-22

ITERATION LIMIT
```

2.3 Min-Max Method

Description of the Basic Method

Extreme difficulties in properly initializing the Newton-Raphson method with slack variables has led to a number of other algorithms for minimization problems which are constrained by algebraic inequalities. One such method is called the min-max method [5] which is based on a number of observations about the Newton-Raphson method itself. These observations are made considering inequalities

$$\mathbf{g(c)} \leq \mathbf{0} \text{ and } \boldsymbol{\lambda} \geq \mathbf{0} \tag{2.16}$$

on constraint functions and the corresponding Lagrange multipliers.

The first observation concerns the set of constraints that is pertinent to any iteration of the selectable vectors $\mathbf{c}$ and $\boldsymbol{\lambda}$ of the Lagrangian. The set of effective inequality constraints is defined by index from

$$I = \left\{ i : g_i(\mathbf{c}) > 0 \text{ or } \lambda_i > 0 \right\} \tag{2.17}$$

which permits iterations to fall outside constraint boundaries and also to ignore constraints that are superfluous at the solution point. An objective function

$$L(\mathbf{c}, \hat{\boldsymbol{\lambda}}) = f(\mathbf{c}) + \hat{\boldsymbol{\lambda}}' \hat{\mathbf{g}}(\mathbf{c}) \tag{2.18}$$

can now be defined in terms of vectors,

$$\hat{\mathbf{g}}(\mathbf{c}) = \left\{ g_i(\mathbf{c}) : i \in I \right\} \text{ and } \hat{\boldsymbol{\lambda}} = \left\{ \lambda_i : i \in I \right\}. \tag{2.19}$$

Function $L(\mathbf{c}, \hat{\boldsymbol{\lambda}})$ is identical to the Lagrangian for this problem whenever the inequalities given in equation (2.16) hold. Note also that the Newton-Raphson direction vector $\mathbf{d}$ for vector $\mathbf{c}$ is given by

$$\mathbf{d} = -L_{cc}^{-1} (L_c + \hat{\mathbf{g}}'_c \hat{\boldsymbol{\delta}}) \tag{2.20}$$

where vector $\hat{\boldsymbol{\delta}}$ is an arbitrary direction vector for vector $\hat{\boldsymbol{\lambda}}$.

Now suppose that a series of Newton-Raphson iterations is embarked upon holding vector $\hat{\boldsymbol{\lambda}}$ fixed so that $\hat{\boldsymbol{\delta}} = \mathbf{0}$. These iterations correspond to the minimization of L with respect to $\mathbf{c}$. At the limit point of these iterations, condition $L_c = \mathbf{0}$ holds. Of course, one or more constraints may be violated at this limit point and elements of vector $\hat{\boldsymbol{\lambda}}$ may have incorrect values.

OPTIMIZATION WITH FUNCTIONS

Next suppose that a series of iterations, not necessarily Newton-Raphson, is embarked upon for the purpose of adjusting vector $\hat{\lambda}$. In particular, vector $\hat{\delta}$ is iterated so that the objective function $L(\mathbf{c} - L_{cc}^{-1} \hat{\mathbf{g}}_c' \hat{\delta}, \hat{\lambda} + \hat{\delta})$ is maximized subject to the inequality constraint

$$\hat{\lambda} + \hat{\delta} \geq \mathbf{0}. \qquad (2.21)$$

A Newton-Raphson direction vector can be used for this maximization problem by introducing slack variables corresponding to inequality constraints given in equation (2.21). Moreover, a successful algorithm for initializing these Newton-Raphson iterations with slack variables has been devised and implemented in the subprogram described below.

The min-max method utilizes the minimization and maximization steps described above in an iterative fashion. The subprogram described below is also embellished by eliminating the need for direct computation of matrix L_{cc}^{-1} in the minimization step. Specifically, the Fletcher-Powell method [1] of generating conjugate direction vectors results in an approximation of matrix L_{cc}^{-1}. This approximation accrues during the iterations of the minimization step in the min-max method using the following method of selecting direction vectors. Define vector

$$\mathbf{e} = -f_c \qquad (2.22)$$

so that direction vector $\mathbf{d}$ can be defined iteratively as

$$\mathbf{d}_i = \Gamma_i \mathbf{e}_i \qquad (2.23)$$

where

$$\Gamma_{i+1} = \Gamma_i + \alpha_i \frac{\mathbf{d}_i \mathbf{d}_i'}{\mathbf{d}_i' \mathbf{f}_i} - \frac{(\Gamma_i \mathbf{f}_i)(\Gamma_i \mathbf{f}_i)'}{(\Gamma_i \mathbf{f}_i)' \mathbf{f}_i} \qquad (2.24)$$

and

$$\mathbf{f}_i = \mathbf{e}_i - \mathbf{e}_{i+1} \qquad (2.25)$$

for $i = 0, 1, \ldots$. Iterations can be initialized by any positive definite matrix and are often accomplished with $\Gamma_0 = \mathbf{I}$. In addition, condition $L_{cc}^{-1} = \Gamma_n$ holds when condition $n = \text{dim}\mathbf{c}$ holds and Lagrangian L is quadratic in vector $\mathbf{c}$. Matrix Γ_i thus is initialized at the beginning of every min-max step so that matrix L_{cc}^{-1} is approximated as well as possible during the minimization step.

MIN3 and MIN3S

The successive approximation procedure outlined previously includes many of the aspects of MIN1 for the minimization step and many of the aspects of MIN2 for the maximization step. This procedure is implemented by MIN3 which is summarized as follows:

MIN3 function: Min (Fletcher-Powell)-max (Newton-Raphson) method of successive approximations for minimizing functions with equality and inequality constraints.

MIN3 dummy argument list: None.

MIN3 input data list: C1/C2/C3/NI,NT,NR,NC,NEC,NIC/RATIO,DFMIN, FMIN,FMAX/ALMIN,ALMAX,RMIN,RMAX/GMAX/CV/LAMV/DDLIM, where
    NI = maximum number of iterations.
    NT = maximum number of trials per iteration.
    NR = number of iterations performed before maximization steps. $1 \leqslant NC \leqslant 21, 0 \leqslant NEC \leqslant 3-NIC, 0 \leqslant NIC \leqslant 3-NEC$.
    RATIO = minimum value of $|s(\alpha)/s(0)|$ requiring additional trials.
    DFMIN = minimum value of $|L_c|$ requiring additional trials and iterations.
    FMIN = lower bound on $f(\mathbf{c})$.
    FMAX = upper bound on $f(\mathbf{c})$.
    ALMIN = minimum initial value of α.
    ALMAX = maximum initial value of α.
    RMIN = minimum fraction of $(\alpha - \beta)$ used in interval reduction (<1).
    RMAX = maximum fraction of $(\alpha - \beta)$ used in extrapolation (>1).
    GMAX = maximum error in satisfying constraints not requiring additional iterations.
    CV(NC) = $\mathbf{c}$.
    LAMV(NEC+NIC) = $\boldsymbol{\lambda}$.
    DDLIM(NC,NC) = positive definite approximation to L_{cc}^{-1}.

MIN3 output data list: CV/LAMV/DDLIM, where
    CV(NC) = $\mathbf{c}$.
    LAMV(NEC+NIC) = $\boldsymbol{\lambda}$.
    DDLIM(NC,NC) = positive definite approximation to L_{cc}^{-1}.

Subprograms called by MIN3: MI and MIN3S for the computation of $F = f(\mathbf{c})$, DFV(NC) = $f_c{'}$, GV(NEC+NIC) = $\mathbf{g}$, and DGM(NEC+NIC,NC) = $\mathbf{g}_c$.

Program MIN3 is listed in Table 2-5. User problems are specified for sub-

Table 2-5
Listing of MIN3

```
 1         SUBROUTINE MIN3
 2   C     REAL*8 DGM(N1,NC),DLIM(NC,NC),BM(N1,NC),FAM(N1,NC),CM(N1,N1),
 3   C    1MM(NT,NT),CV(NC),LAMV(N1),DFV(NC),GV(N1),DLV(NC),DSV(NC),
 4   C    2SCV(NC),AV(N1),DFAV(N1),D1V(N1),D2V(NIC),D3V(NIC),DD1V(N1),
 5   C    3D1FV(NT),DDFV(NT),DCV(NC)
 6   C     LOGICAL*1 IV(N1)
 7         REAL*8 DGM(3,21),DLIM(21,21),BM(3,21),FAM(3,21),CM(3,3),
 8        1MM(9,9),CV(21),LAMV(3),DFV(21),GV(3),DLV(21),DSV(21),
 9        2SCV(21),AV(3),DFAV(3),D1V(3),D2V(3),D3V(3),DD1V(3),
10        3D1FV(9),DDFV(9),DCV(21)
11         LOGICAL*1 IV(3)
12         REAL*8 A,ALMIN,ALMAX,B,C,D,DC,DF,DFA,DFB,DFMIN,D1,D2,D3,E,F,
13        1FA,FB,FMIN,FMAX,GMAX,L,RATIO,RMIN,RMAX,SDF,X
14         REAL*8 DABS,DSQRT,DMAX1,DMIN1
15         DATA MC/21/,M1/3/,MT/9/
16         DATA F/0.D0/,DFV/21*0.D0/,GV/3*0.D0/,DGM/63*0.D0/
17         LOGICAL*1 COM(65)
18   1000  FORMAT(65A1)
19   1001  FORMAT(4X,65A1)
20   1002  FORMAT(4D18.10)
21   1003  FORMAT(6I10)
22   1004  FORMAT(4X,4D18.10)
23   1005  FORMAT(' 1')
24   1006  FORMAT(//,'      NI,NT,NR,NC,NEC,NIC')
25   1007  FORMAT('       GMAX')
```

(continued)

Table 2-5 Continued

```
26       1008 FORMAT(' RATIO,DFMIN,FMIN,FMAX')
27       1009 FORMAT(' ALMIN,ALMAX,RMIN,RMAX')
28       1010 FORMAT(' CV')
29       1011 FORMAT(' LAMV')
30       1012 FORMAT(' DDLIM')
31       1013 FORMAT(//,'        ITERATION=',I2,' TRIAL=',I2,'    F=',
32            1D18.10,' L=',D18.10)
33       1014 FORMAT(' ALPHA=',D18.10,'   DL=',D18.10,'   |DLV|=',D18.10)
34       1015 FORMAT(//,' RESET CV')
35       1016 FORMAT(//,' DESIGN COMPLETE')
36       1017 FORMAT(//,' TRIAL LIMIT')
37       1018 FORMAT('0       ITERATION=',I2,' TRIAL=',I2,'    |DLV|=',D18.10)
38       1019 FORMAT(' DLAMV')
39       1020 FORMAT(//,' ITERATION LIMIT')
40       1021 FORMAT(' GV')
41            WRITE (6,1005)
42            DO 1022 I=1,3
43            READ (5,1000) COM
44       1022 WRITE (6,1001) COM
45            READ (5,1003) NI,NT,NR,NC,NEC,NIC
46            READ (5,1002) RATIO,DFMIN,FMIN,FMAX
47            READ (5,1002) ALMIN,ALMAX,RMIN,RMAX
48            N1=NEC+NIC
49            IF (N1) 1024,1024,1023
50       1023 READ (5,1002) GMAX
51       1024 READ (5,1002) (CV(I),I=1,NC)
```

```
52        WRITE (6,1006)
53        WRITE (6,1003)         NI,NT,NR,NC,NEC,NIC
54        WRITE (6,1008)
55        WRITE (6,1004)         RATIO,DFMIN,FMAX
56        WRITE (6,1009)
57        WRITE (6,1004)         ALMIN,ALMAX,RMIN,RMAX
58        IF (N1) 1026,1026,1025
59   1025 WRITE (6,1007)
60        WRITE (6,1004)         GMAX
61        READ (5,1002)          (LAMV(I),I=1,N1)
62   1026 READ (5,1002)          ((DLIM(I,J),I=1,J),J=1,NC)
63        WRITE (6,1012)
64        WRITE (6,1004)         ((DLIM(I,J),I=1,J),J=1,NC)
65        NCC=0
66        NTC=0
67        C=0.D0
68        X=1.D0-1.D0/RMAX
69        NRC=NR-1
70        CALL MIN3S(MC,M1,CV,IV,F,DFV,GV,DGM,1)
71        CALL MIN3S(MC,M1,CV,IV,F,DFV,GV,DGM,2)
72        L=F
73        DO 1027 I=1,NC
74        SCV(I)=CV(I)
75        DLV(I)=DFV(I)
76        DO 1027 J=I,NC
77   1027 DLIM(J,I)=DLIM(I,J)
78        IF (N1) 1033,1033,1028
```

OPTIMIZATION WITH FUNCTIONS   47

(continued)

Table 2-5 Continued

```
79          DO 1028 I=1,N1
80  1028    DO 1029 I=1,N1
81  1029    I=L+LAMV(I)*GV(I)
82          DO 1030 I=1,N1
83  1030    IV(I)=.TRUE.
84          DO 1032 I=1,NC
85          D=DLV(I)
86          DO 1031 J=1,N1
87  1031    D=D+DGM(J,I)*LAMV(J)
88  1032    DLV(I)=D
89  1033    DF=0.D0
90          DO 1035 I=1,NC
91          DSV(I)=DLV(I)
92          D=0.D0
93          DO 1034 J=1,NC
94  1034    D=D-DLIM(I,J)*DLV(J)
95          DCV(I)=D
96  1035    DF=DF+D*DLV(I)
97          SDF=RATIO*DABS(DF)
98          D3=DSQRT(-DF)
99          WRITE (6,1013) NCC,NTC,F,L
100         WRITE (6,1014) C,DF,D3
101         WRITE (6,1010)
102         WRITE (6,1004) (CV(I),I=1,NC)
103         IF (N1) 1037,1037,1036
104 1036    WRITE (6,1021)
            WRITE (6,1004) (GV(I),I=1,N1)
```

```
105        WRITE (6,1011)
106        WRITE (6,1004) (LAMV(I),I=1,N1)
107  1037  IF (FMAX .GT. F) GO TO 1038
108        WRITE (6,1015)
109        RETURN
110  1038  IF (DFMIN .LT. D3) GO TO 1039
111        GO TO 1055
112  1039  DC=2.D0*(FMIN-L)/DF
113  1040  DC=DMAX1(ALMIN,DMIN1(DC,ALMAX))
114  1041  NTC=NTC+1
115        A=C
116        FA=L
117        DFA=DF
118        C=A+DC
119        DO 1042 I=1,NC
120  1042  CV(I)=SCV(I)+C*DCV(I)
121        CALL MIN3S(MC,M1,CV,IV,F,DFV,GV,DGM,2)
122        L=F
123        DO 1043 I=1,NC
124  1043  DLV(I)=DFV(I)
125        IF (N1) 1048,1048,1044
126  1044  DO 1045 I=1,N1
127  1045  L=L+LAMV(I)*GV(I)
128        DO 1047 I=1,NC
129        D=DLV(I)
130        DO 1046 J=1,N1
```

(continued)

Table 2-5 Continued

```
131  1046  D=D+DGM(J,I)*LAMV(J)
132  1047  DLV(I)=D
133  1048  DF=0.D0
134        D1=0.D0
135        DO 1049 I=1,NC
136        DF=DF+DCV(I)*DLV(I)
137  1049  D1=D1+DLV(I)*DLV(I)
138        D2=DSQRT(D1)
139        IF (SDF .GE. DABS(DF)) GO TO 1080
140        WRITE (6,1013) NCC,NTC,F,L
141        WRITE (6,1014) C,DF,D2
142        WRITE (6,1010)
143        WRITE (6,1004) (CV(I),I=1,NC)
144        IF (N1) 1051,1051,1050
145  1050  WRITE (6,1021)
146        WRITE (6,1004) (GV(I),I=1,N1)
147  1051  IF (FMAX .GT. F) GO TO 1054
148  1052  IF (NT .GT. NTC) GO TO 1053
149        GO TO 1063
150  1053  DC=RMIN*DC
151        C=A
152        L=FA
153        DF=DFA
154        GO TO 1041
155  1054  IF (DFMIN .LT. D2) GO TO 1060
156  1055  NRC=NCC
```

```
157            IF (N1) 1059,1059,1056
158       1056 DO 1058 I=1,N1
159            IF (I .GT. NEC) GO TO 1057
160            IF (DABS(GV(I)) .GT. GMAX) GO TO 1080
161            GO TO 1058
162       1057 IF (GV(I) .GT. GMAX) GO TO 1080
163       1058 CONTINUE
164       1059 WRITE (6,1016)
165            GO TO 1144
166       1060 IF (DF) 1061,1061,1064
167       1061 IF (NT .LE. NTC) GO TO 1063
168            IF (X*DFA .LE. DF) GO TO 1062
169            DC=DC*RMAX
170            GO TO 1041
171       1062 DC=DC*DFA/(DFA-DF)
172            GO TO 1041
173       1063 WRITE (6,1017)
174            GO TO 1144
175       1064 B=C
176            FB=L
177            DFB=DF
178       1065 D1=DFA+DFB+3.D0*(FA-FB)/DC
179            L2=2.D0*(DFA+D1)
180            D3=DFA+DFB+2.D0*D1
181            IF (1.D-5*DABS(D2) .LE. DABS(D3)) GO TO 1066
182            C=B-(DFA+2.D0*D1)*DC/D2
183            GO TO 1067
```

(continued)

Table 2-5 Continued

```
184     1066    D2=DSQRT(D1**2-DFA*DFB)
185             C=B-(DFB+D1-D2)*DC/D3
186     1067    DO 1068 I=1,NC
187     1068    CV(I)=SCV(I)+C*DCV(I)
188             CALL MIN3S(MC,M1,CV,IV,F,DFV,GV,DGM,2)
189             L=F
190             DO 1069 I=1,NC
191     1069    DLV(I)=DFV(I)
192             IF (N1) 1074,1074,1070
193     1070    DO 1072 I=1,NC
194             D=DLV(I)
195             DO 1071 J=1,N1
196     1071    D=D+DGM(J,I)*LAMV(J)
197     1072    DLV(I)=D
198             DO 1073 J=1,N1
199     1073    L=L+LAMV(J)*GV(J)
200     1074    LF=0.D0
201             D1=0.D0
202             DO 1075 I=1,NC
203             DF=DF+DCV(I)*DLV(I)
204     1075    D1=D1+DLV(I)*DLV(I)
205             D2=DSQRT(D1)
206             IF (SDF .GE. DABS(DF)) GO TO 1080
207             NTC=NTC+1
208             WRITE (6,1013) NCC,NTC,F,L
209             WRITE (6,1014) C,DF,D2
```

```
210         WRITE (6,1010)
211         WRITE (6,1004)    (CV(I),I=1,NC)
212         IF (N1) 1077,1077,1076
213  1076   WRITE (6,1021)
214         WRITE (6,1004)    (GV(I),I=1,N1)
215  1077   IF (FMAX .LE. F) GO TO 1052
216         IF (NT .LE. NTC) GO TO 1063
217         IF (DF) 1079,1078,1078
218  1078   DC=C-A
219         GO TO 1064
220  1079   A=C
221         FA=L
222         DFA=DF
223         DC=B-C
224         GO TO 1065
225  1080   DO 1081 I=1,NC
226  1081   SCV(I)=DLV(I)-DSV(I)
227         E=0.D0
228         DO 1083 I=1,NC
229         D=0.D0
230         DO 1082 J=1,NC
231  1082   D=D+DLIM(I,J)*SCV(J)
232         DSV(I)=D
233  1083   E=E+D*SCV(I)
234         E=1.D0/E
235         D=0.D0
236         DO 1084 I=1,NC
237         DCV(I)=C*DCV(I)
```

(continued)

Table 2-5 Continued

```
238   1084  D=D+DCV(I)*SCV(I)
239         D=1.D0/D
240         DO 1085 I=1,NC
241         DO 1085 J=I,NC
242         DLIM(I,J)=DLIM(I,J)-DSV(I)*DSV(J)*E+DCV(I)*DCV(J)*D
243   1085  DLIM(J,I)=DLIM(I,J)
244         WRITE (6,1012)
245         WRITE (6,1004) ((DLIM(I,J),I=1,J),J=1,NC)
246         IF (NRC .GT. NCC) GO TO 1139
247         NRC=NRC+NR
248         IF (N1) 1139,1139,1086
249   1086  N=NEC
250         IF (NIC) 1092,1092,1087
251   1087  DO 1091 I=1,NIC
252         IF (N .EQ. NC) GO TO 1092
253         J=I+NEC
254         IF (GV(J)) 1088,1088,1089
255   1088  IF (LAMV(J)) 1090,1090,1089
256   1089  IV(J)=.TRUE.
257         N=N+1
258         GO TO 1091
259   1090  IV(J)=.FALSE.
260   1091  CONTINUE
261   1092  IF (N) 1139,1139,1093
262   1093  M=N-NEC
263         K=0
```

```
264        CALL MIN3S(MC,M1,CV,IV,F,DFV,GV,DGM,3)
265        DO 1095 I=1,N1
266        IF (.NOT. IV(I)) GO TO 1095
267        K=K+1
268        AV(K)=-GV(I)
269        DFAV(K)=LAMV(I)
270        DO 1094 J=1,NC
271   1094 BM(K,J)=DGM(I,J)
272   1095 CONTINUE
273        DO 1097 I=1,N
274        DO 1097 J=1,NC
275        D=0.D0
276        DO 1096 K=1,NC
277   1096 D=D+BM(I,K)*DLIM(K,J)
278   1097 FAM(I,J)=D
279        DO 1099 I=1,N
280        D=AV(I)
281        DO 1098 J=1,NC
282   1098 D=D+FAM(I,J)*DLV(J)
283   1099 AV(I)=D
284        DO 1101 I=1,N
285        DO 1101 J=I,N
286        D=0.D0
287        DO 1100 K=1,NC
288   1100 D=D+FAM(I,K)*BM(J,K)
289        MM(I,J)=D
290        MM(J,I)=D
```

Table 2-5 Continued

```
291          CM(I,J)=D
292     1101 CM(J,I)=D
293          CALL MI(MT,N,MM)
294          SDF=0.D0
295          DO 1104 I=1,N
296          D=0.D0
297          DO 1102 J=1,N
298     1102 D=D-MM(I,J)*AV(J)
299          IF(I.GT.NEC) GO TO 1103
300          D1V(I)=D
301          GO TO 1104
302     1103 D1V(I)=DMAX1(-DFAV(I),D)
303     1104 SDF=SDF+AV(I)**2
304          SDF=RATIO*DSQRT(SDF)
305          IF(M) 1108,1108,1105
306     1105 DO 1107 I=1,M
307          K=I+NEC
308          D2V(I)=DSQRT(2.D0*(DFAV(K)+D1V(K)))
309          D=AV(K)
310          DO 1106 J=1,N
311     1106 D=D+CM(K,J)*D1V(J)
312     1107 D3V(I)=D
313     1108 NS=N+M
314          MS=2*M
315          NT=NS+M
316          DO 1109 I=1,NT
```

OPTIMIZATION WITH FUNCTIONS

```
317       1109 D1FV(I)=0.D0
318       1110 DO 1111 I=1,N
319       1111 D1V(I)=D1V(I)+D1FV(I)
320            IF (M) 1114,1114,1112
321       1112 DO 1113 I=1,M
322            D2V(I)=D2V(I)+D1FV(I+N)
323       1113 D3V(I)=D3V(I)+D1FV(I+NS)
324       1114 DO 1117 I=1,N
325            D=AV(I)
326            IF (I .LE. NEC) GO TO 1115
327            D=D-D3V(I-NEC)
328       1115 DO 1116 J=1,N
329       1116 D=D+CM(I,J)*D1V(J)
330       1117 D1FV(I)=D
331            IF (M) 1120,1120,1118
332       1118 DO 1119 I=1,M
333            D1FV(I+N)=D3V(I)*D2V(I)
334       1119 D1FV(I+NS)=5.D-1*D2V(I)**2-(DFAV(I+NEC)+D1V(I+NEC))
335       1120 DFB=0.D0
336            DO 1121 I=1,NT
337       1121 DFB=DFB+D1FV(I)**2
338            DFB=DSQRT(DFB)
339            WRITE (6,1018) NCC,NTC,DFB
340            WRITE (6,1019)
341            WRITE (6,1004) (D1V(I),I=1,N)
342            IF (DFMIN .GE. DFB) GO TO 1131
343            IF (DFB .LE. SDF) GO TO 1131
344            IF (NTC .GE. NT) GO TO 1063
```

(continued)

Table 2-5 Continued

```
345              NTC=NTC+1
346              DO 1128 J=1,NT
347              DDFV(J)=D1FV(J)
348              IF (J .GT. N) GO TO 1123
349              DO 1122 I=1,J
350       1122   MM(I,J)=CM(I,J)
351              GO TO 1127
352       1123   IF (J .GT. NS) GO TO 1125
353              DO 1124 I=1,J
354       1124   MM(I,J)=0.D0
355              MM(J,J)=D3V(J-N)
356              GO TO 1127
357       1125   DO 1126 I=1,J
358       1126   MM(I,J)=0.D0
359              MM(J-MS,J)=-1.D0
360              MM(J-M,J)=D2V(J-NS)
361       1127   DO 1128 I=1,J
362       1128   MM(J,I)=MM(I,J)
363              CALL MI(MT,NT,MM)
364              DO 1130 I=1,NT
365              D=0.D0
366              DO 1129 J=1,NT
367       1129   D=D-MM(I,J)*DDFV(J)
368       1130   D1FV(I)=D
369              GO TO 1110
370       1131   DO 1133 I=1,N
```

```
        IF (I .GT. NEC) GO TO 1132
        DD1V(I)=D1V(I)
        GO TO 1133
1132    DD1V(I)=DMAX1(-DFAV(I),D1V(I))
1133    CONTINUE
        K=0
        DO 1136 I=1,N1
        IF (IV(I)) GO TO 1134
        D1V(I)=0.D0
        GO TO 1135
1134    K=K+1
        D1V(I)=DD1V(K)
1135    LAMV(I)=LAMV(I)+D1V(I)
1136    L=L+D1V(I)*GV(I)
        DO 1138 I=1,NC
        D=DLV(I)
        DO 1137 J=1,N1
1137    D=D+DGM(J,I)*D1V(J)
1138    DLV(I)=D
1139    DF=0.D0
        D2=0.D0
        DO 1141 I=1,NC
        DSV(I)=DLV(I)
        SCV(I)=CV(I)
        D=0.D0
        DO 1140 J=1,NC
1140    D=D-DLIM(I,J)*DLV(J)
```

Table 2-5 Continued

```
398          DCV(I)=D
399          D2=D2+DLV(I)**2
400  1141    DF=DF+D*DLV(I)
401          SDF=RATIO*DABS(DF)
402          DC=C
403          C=0.D0
404          NTC=0
405          NCC=NCC+1
406          D2=DSQRT(D2)
407          WRITE (6,1013) NCC,NTC,F,L
408          WRITE (6,1014) C,DF,D2
409          WRITE (6,1010)
410          WRITE (6,1004) (CV(I),I=1,NC)
411          IF (N1) 1143,1143,1142
412  1142    WRITE (6,1021)
413          WRITE (6,1004) (GV(I),I=1,N1)
414          WRITE (6,1011)
415          WRITE (6,1004) (LAMV(I),I=1,N1)
416  1143    IF (NI .GT. NCC) GO TO 1040
417          WRITE (6,1020)
418  1144    WRITE (6,1005)
419          WRITE (7,1002) (SCV(I),I=1,NC)
420          IF (N1) 1146,1146,1145
421  1145    WRITE (7,1002) (LAMV(I),I=1,N1)
422  1146    WRITE (7,1002) ((DLIM(I,J),I=1,J),J=1,NC)
423          RETURN
424          END
```

OPTIMIZATION WITH FUNCTIONS

program MIN3 by writing an appropriate subprogram MIN3S. Specifications for writing this subprogram are given as follows:

MIN3S function: Compute $f(\mathbf{c}), f_\mathbf{c}, \mathbf{g(c)}$, and $\mathbf{g_c}$ for use with MIN3.

MIN3S dummy argument list: MC,MG,CV,IV,F,DFV,GV,DGM,MODE, where
   CV(MC) = **c**.
   IV(MG) = logical vector of equality and effective inequality (T) and ineffective inequality (F) constraints for computing DGM.
   F = $f(\mathbf{c})$.
   DFV(MC) = $f_\mathbf{c}$.
   GV(MG) = **g(c)**.
   DGM(MG,MC) = $\mathbf{g_c}$.
   MODE = 1 for program initialization, 2 for the computation of F,DFV,GV, and 3 for the computation of DGM.

Example 2B

The constrained minimization problem given in Example 2A can also be solved using the min-max procedure which is implemented by MIN3. Correspondingly, user-supplied calling program MIN3C and subprogram MIN3S are listed in Table 2-6a, and user-supplied input data are listed in Table 2-6b. Finally, program printed output is listed in Table 2-6c. The min-max method is seen to yield results which are comparable to those of the Newton-Raphson method for this example.

Table 2-6a
Listing of User-supplied Programs for Example 2B

```
 1      C      MIN3C    EXAMPLE 2B
 2             CALL MIN3
 3             STOP
 4             END
 5             SUBROUTINE MIN3S(MC,M1,CV,IV,F,DFV,GV,DGM,MODE)
 6             REAL*8 DGM(M1,MC),CV(MC),DFV(MC),GV(M1),F
 7             LOGICAL*1 IV(M1)
 8             REAL*8 FV(2),D
 9             DATA FV/1.D0,2.D0/
10             GO TO (1000,1001,1004),MODE
11      1000   DGM(1,1)=1.D0
12             DGM(2,1)=1.D0
13             DGM(1,2)=1.D0
14             DGM(2,2)=2.D0
15             DGM(1,3)=1.D0
16             DGM(2,3)=1.D0
17             RETURN
18      1001   F=CV(1)**2+2.D0*CV(2)**2+3.D0*CV(3)**2
19             DFV(1)=2.D0*CV(1)
20             DFV(2)=4.D0*CV(2)
21             DFV(3)=6.D0*CV(3)
22             DO 1003 I=1,2
23             D=FV(I)
24             DO 1002 J=1,3
25      1002   D=D+DGM(I,J)*CV(J)
26      1003   GV(I)=D
27      1004   RETURN
28             END
```

Table 2-6b
Listing of User-supplied Input Data for Example 2B

```
 1  MIN3(MI,MIN3S)   EXAMPLE 2B
 2  MERRIAM
 3  UNIVERSITY OF ROCHESTER
 4           3           8           1           3           2
 5    .01                                       0.            30.
 6    .01                           10.          0.5          10.
 7    .000001        .000001
 8   3.             -3.             0.
 9   0.              0.             0.           0.25         0.
10   0.5             0.            .16667
11   0.
```

Table 2-6c
Printed Output from MIN3 for Example 2B

```
MIN3(MI,MIN3S)    EXAMPLE 2B
MERRIAM
UNIVERSITY OF ROCHESTER


NI,NT,NR,NC,NEC,NIC
  3      8      1      3      0      2
RATIO,DFMIN,FMIN,FMAX
  0.1000000000D-01  0.1000000000D-05  0.0000000000D 00  0.3000000000D 02
ALMIN,ALMAX,RMIN,RMAX
  0.1000000000D-01  0.1000000000D 02  0.5000000000D 00  0.1000000000D 02
GMAX
  0.1000000000D-05
DDLIM
  0.5000000000D 00  0.0000000000D 00  0.2500000000D 00  0.0000000000D 00
  0.0000000000D 00  0.1666700000D 00


ITERATION= 0   TRIAL= 0     F=  0.2700000000D 02    L=  0.2700000000D 02
ALPHA=  0.0000000000D 00   DL= -0.5400000000D 02   |DLV|=  0.7348469228D 01
CV
  0.3000000000D 01 -0.3000000000D 01  0.0000000000D 00
GV
  0.1000000000D 01 -0.1000000000D 01
```

OPTIMIZATION WITH FUNCTIONS

```
LAMV
   0.000000000D 00  0.000000000D 00
DDLIM
   0.500000000D 00  0.000000000D 00  0.250000000D 00  0.000000000D 00
   0.000000000D 00  0.166670000D 00
ITERATION= 0  TRIAL= 1  |DLV|= 0.1319936429D 02
DLAMV
   0.000000000D 00  0.400000000D 01
ITERATION= 0  TRIAL= 2  |DLV|= 0.2549077030D 01
DLAMV
   0.000000000D 00  0.1925924609D 01
ITERATION= 0  TRIAL= 3  |DLV|= 0.4056641467D 00
DLAMV
   0.000000000D 00  0.1341165440D 01
ITERATION= 0  TRIAL= 4  |DLV|= 0.2868303071D-01
DLAMV
   0.000000000D 00  0.1210583082D 01
ITERATION= 0  TRIAL= 5  |DLV|= 0.2192063301D-03
DLAMV
   0.000000000D 00  0.1200078007D 01

ITERATION= 1  TRIAL= 0        F= 0.4437342592D-30   L= 0.2400156015D 01
ALPHA= 0.000000000D 00     DL= -0.2400316840D 01  |DLV|= 0.2939578769D 01
CV
  -0.2220446049D-15  -0.4440892099D-15   0.000000000D 00
GV
   0.1000000000D 01   0.2000000000D 01
```

(continued)

Table 2-6c Continued

```
LAMV
  0.000000000D 00   0.1200078007D 01
DDLIM
  0.5000003000D 00   0.3000090000D-06   0.2500003000D 00  -0.9000089997D-06
 -0.9000089997D-06   0.1666693666D 00
ITERATION= 1   TRIAL= 1   |DLV|=   0.2220446052D-15
DLAMV
 -0.7800727047D-04

ITERATION= 2   TRIAL= 0         F=  0.1200160820D 01      L=  0.1200000005D 01
ALPHA=  0.0000000000D 00   DL= -0.1086200318D-07   |DLV|=  0.2020679904D-03
CV
 -0.6000390036D 00  -0.6000390036D 00  -0.2000170015D 00
GV
 -0.4000950087D 00  -0.1340123775D-03
LAMV
  0.0000000000D 00   0.1200000000D 01
DDLIM
  0.5000004752D 00   0.4751907542D-06   0.2500004752D 00  -0.1090145371D-05
 -0.1090145371D-05   0.1666691676D 00
ITERATION= 2   TRIAL= 1   |DLV|=   0.1527622574D-30
DLAMV
  0.1060463898D-14
```

```
ITERATION=  3    TRIAL=  0     F=   0.1200000000D 01    L=  0.1200000000D 01
ALPHA=   0.0000000000D 00     DL= -0.2799366311D-19   |DLV| = 0.4006172965D-09
CV
  -0.6000000000D 00  -0.6000000000D 00  -0.1999999999D 00
GV
  -0.4000000000D 00  -0.5551115123D-16
LAMV
   0.0000000000D 00   0.1200000000D 01

ITERATION LIMIT
```

3
Formulation of Performance Functionals

The computer programs presented in this chapter are used in formulating performance functionals for the design of controls for linear systems with deterministic signals. These design techniques are based on excess-pole specifications which can also be used to synthesize a matrix of closed-loop transfer functions in somewhat restricted circumstances [6].

3.1 Excess-Pole-Specification Synthesis Method

Description of the Basic Method

The notion of an excess-pole specification can be explained in terms of performance integral

$$I = \int_0^\infty \left\{ \frac{1}{2} e'(t)e(t) \right\} dt, \qquad (3.1)$$

where $e(t)$ denotes an error vector which is defined in terms of the open-loop linear system

$$\dot{x} = Fx + Gu, \, x(0) = x_o, \qquad (3.2)$$

and

$$y = Hx. \qquad (3.3)$$

Specifically, this error vector is defined by

$$e = (c - y_o) - \left| \sum_{k=0}^{K} P_k \frac{d^k(y - y_o)}{dt^k} \right., \qquad (3.4)$$

where c is an arbitrary deterministic command vector and

$$y_o = Hx_o. \qquad (3.5)$$

The matrix of polynomials

$$P(s) = \sum_{k=0}^{K} P_k s^k \qquad (3.6)$$

is called the excess-pole specification and assumes a special significance whenever condition $I = 0$ holds for all initial states.

Appropriate algebraic manipulation permits rewriting (3.4) in terms of state and control vectors as

$$e = y_c - Dx - \sum_{\ell=0}^{K-1} E_\ell \frac{d^\ell u}{dt^\ell}, \qquad (3.7)$$

where

$$y_c = c + (P_0 - I)y_o. \qquad (3.8)$$

Matrices D and E_ℓ are polynomials in matrix F^k with coefficients P_k. The above expression for error vector $e(t)$ further simplifies when the excess-pole specification is selected properly so that condition $E_\ell = 0$ holds for $\ell = 1, 2, \ldots, (K-1)$ and matrix E_0 is nonsingular. In particular, selection of control vector

$$u = E_0^{-1}(y_c - Dx) \qquad (3.9)$$

results in condition $e(t) = 0$ for all command vectors and initial states. This particular set of circumstances results in the synthesis of a closed-loop matrix of transfer functions which corresponds to the excess-pole specification in a precise manner.

Even when condition $E_\ell = 0$ holds for $\ell = 1, 2, \ldots, (K-1)$ and matrix E_0 is nonsingular, feedback control corresponding to equation (3.9) generally cannot be implemented because the entire state vector cannot be measured. In this situation, the error vector is not identically zero and performance integral

$$I = \int_0^\infty \left\{ \frac{1}{2}(y_c - Dx - E_0 u)'(y_c - Dx - E_0 u) \right\} dt \qquad (3.10)$$

does not vanish. However, optimization techniques can be used to select the parameters of an implementable feedback controller that minimizes the performance integral given in equation (3.10).

SYN

The matrices appearing in equation (3.7) are computed in program SYN. The coefficient matrices of the closed-loop system, which corresponds to minimizing the performance integral given in equation (3.10) with respect to control vector **u**, are also computed in program SYN. This program is summarized as follows:

SYN function: Compute

$$\mathbf{D} = \sum_{k=0}^{K} \mathbf{P}_k \mathbf{H} \mathbf{F}^k ,$$

$$\mathbf{E}_\ell = \left[\sum_{k=0}^{K-1-\ell} \mathbf{P}_{k+1+\ell} \mathbf{H} \mathbf{F}^k \right] \mathbf{G} \text{ for } \ell = 0, 1, \ldots, (K-1) ,$$

$$\mathbf{K} = (\mathbf{E}_0' \mathbf{E}_0)^{-1} \mathbf{E}_0' \mathbf{D} ,$$

$$\mathbf{A} = \mathbf{F} - \mathbf{G}\mathbf{K}$$

and

$$\mathbf{B} = \mathbf{G}(\mathbf{E}_0' \mathbf{E}_0)^{-1} \mathbf{E}_0' .$$

SYN dummy argument list. None.

SYN input data list: C1/C2/C3/NU,NX,NY,ND/FM/GM/HM/PA, where
$1 \leqslant NU \leqslant 3, 1 \leqslant NX \leqslant 7, 1 \leqslant NY \leqslant 7$.
FM(NX,NX) = **F**.
GM(NX,NU) = **G**.
HM(NY,NX) = **H**.
PA(ND,NY,NY) = $\mathbf{P}_0/\mathbf{P}_1/\ldots$.

SYN output data list: KM/AM/BM, where
KM(NU,NX) = **K**.
AM(NX,NX) = **A**.
BM(NX,NU) = **B**.

Subprograms called by SYN: MI.

Program SYN is listed in Table 3-1. There are no user-supplied programs required for program SYN because user's problems are specified entirely by input data.

Example 3A

An elementary example of the use of program SYN is the computation of closed-loop coefficient matrices corresponding to

$$\mathbf{F} = \begin{bmatrix} 0 & 0 & 0 \\ 1 & -3 & -1 \\ 0 & 1 & -1 \end{bmatrix}, \mathbf{G} = \begin{bmatrix} 5 \\ 0 \\ 0 \end{bmatrix}, \mathbf{H} = [0 \ 1 \ 0],$$

and excess-pole specification

$$\mathbf{P}_1 = [1], \mathbf{P}_2 = [2], \text{ and } \mathbf{P}_3 = [1].$$

A listing of program input cards for this example is given in Table 3-2a, and program printed output is listed in Table 3-2b for this example.

3.2 Evaluation of Desired Transfer-Function Matrix

Once closed-loop coefficient matrices have been computed for a given excess-pole specification, stability of the closed-loop system is often of interest as well as the closed-loop matrix of transfer functions. The program described subsequently can be used for these purposes as well as for analyzing any linear constant coefficient system.

Description of the Basic Method

Computation of the matrix of transfer functions $\mathbf{C}(s\mathbf{I} - \mathbf{A})^{-1}\mathbf{B}$ is accomplished using the relationship

$$(s\mathbf{I} - \mathbf{A})^{-1} = \frac{1}{p(s)} \sum_{k=0}^{n-1} \mathbf{\Delta}_k \, s^{(n-1-k)}, \qquad (3.11)$$

Table 3-1
Listing of SYN

```
 1    C     SYN
 2    C     REAL*8DM(NY,NX),EM(NY,NU),FM(NX,NX),GM(NX,NU),HM(NY,NX),
 3    C    1KM(NU,NX),PA(ND,NY,NY),DDM(NY,NX),DEM(NU,NU),DFM(NU,NY),D
 4          REAL*8 DM(7,7),EM(7,3),FM(7,7),GM(7,7),HM(7,7),
 5         1KM(3,7),PA(8,7,7),DDM(7,7),DEM(3,3),DFM(3,7)
 6          DATA MU/3/
 7          REAL*8 D
 8          LOGICAL*1 COM(65)
 9    1000  FORMAT(65A1)
10    1001  FORMAT(4X,65A1)
11    1002  FORMAT(4D18.10)
12    1003  FORMAT(4I10)
13    1004  FORMAT(4X,4D18.10)
14    1005  FORMAT(' 1')
15    1006  FORMAT(//,'              NU,NX,NY,ND')
16    1007  FORMAT('           FM')
17    1008  FORMAT('           GM')
18    1009  FORMAT('           HM')
19    1010  FORMAT(//)
20    1011  FORMAT('           PM(',I2,')')
21    1012  FORMAT('           DM')
22    1013  FORMAT('           EM(',I2,')')
23    1014  FORMAT(//,'           KM')
24    1015  FORMAT('           AM')
25    1016  FORMAT('           BM')
```

(continued)

Table 3-1 Continued

```
26          WRITE (6,1005)
27          DO 1017 I=1,3
28          READ (5,1000) COM
29    1017  WRITE (6,1001) COM
30          READ (5,1003) NU,NX,NY,ND
31          READ (5,1002) ((FM(I,J),I=1,NX),J=1,NX)
32          READ (5,1002) ((GM(I,J),I=1,NX),J=1,NU)
33          READ (5,1002) ((HM(I,J),I=1,NY),J=1,NX)
34          DO 1018 K=1,ND
35    1018  READ (5,1002) ((PA(K,I,J),I=1,NY),J=1,NY)
36          WRITE (6,1006)
37          WRITE (6,1003) NU,NX,NY,ND
38          WRITE (6,1007)
39          WRITE (6,1004) ((FM(I,J),I=1,NX),J=1,NX)
40          WRITE (6,1008)
41          WRITE (6,1004) ((GM(I,J),I=1,NX),J=1,NU)
42          WRITE (6,1009)
43          WRITE (6,1004) ((HM(I,J),I=1,NY),J=1,NX)
44          DO 1019 K=1,ND
45          WRITE (6,1011) K
46    1019  WRITE (6,1004) ((PA(K,I,J),I=1,NY),J=1,NY)
47          WRITE (6,1010)
48          N1=ND-1
49          N2=ND+1
50          MD=0
51          ML=N1-1
```

```
52      1020  L=N1-ML
53      1021  DO 1022 I=1,NY
54            DO 1022 J=1,NX
55      1022  DM(I,J)=0.D0
56            DO 1026 M=1,L
57            N3=N2-M
58            DO 1025 I=1,NY
59            DO 1025 J=1,NX
60            L=0.D0
61            DO 1023 K=1,NX
62      1023  D=D+DM(I,K)*FM(K,J)
63            DO 1024 K=1,NY
64      1024  D=D+PA(N3,I,K)*HM(K,J)
65      1025  DDM(I,J)=D
66            DO 1026 I=1,NY
67            DO 1026 J=1,NX
68      1026  DM(I,J)=DDM(I,J)
69            IF (MD .EQ. 1) GO TO 1031
70            DO 1028 I=1,NY
71            DO 1028 J=1,NU
72            D=0.D0
73            DO 1027 K=1,NX
74      1027  D=D+DM(I,K)*GM(K,J)
75      1028  EM(I,J)=D
76            WRITE (6,1013) ML
77            WRITE (6,1004) ((EM(I,J),I=1,NY),J=1,NU)
78            IF (ML) 1030,1030,1029
```

(continued)

Table 3-1 Continued

```
 79          ML=ML-1
 80  1029    GO TO 1020
 81  1030    MD=1
 82          L=ND
 83          GO TO 1021
 84  1031    WRITE (6,1012)
 85          WRITE (6,1004)  ((DM(I,J),I=1,NY),J=1,NX)
 86          DO 1033 I=1,NU
 87          DO 1033 J=1,NU
 88          D=0.D0
 89          DO 1032 K=1,NY
 90  1032    D=D+EM(K,I)*EM(K,J)
 91          DEM(I,J)=D
 92  1033    DEM(J,I)=D
 93          CALL MI(MU,NU,DEM)
 94          DO 1035 I=1,NU
 95          DO 1035 J=1,NY
 96          D=0.D0
 97          DO 1034 K=1,NU
 98  1034    D=D+DEM(I,K)*EM(J,K)
 99  1035    DFM(I,J)=D
100          DO 1037 I=1,NU
101          DO 1037 J=1,NX
102          D=0.D0
103          DO 1036 K=1,NY
104  1036    D=D+DFM(I,K)*DM(K,J)
```

```
105  1037  KM(I,J)=D
106        DO 1039 I=1,NX
107        DO 1039 J=1,NX
108        D=FM(I,J)
109        DO 1038 K=1,NU
110  1038  D=D-GM(I,K)*KM(K,J)
111  1039  FM(I,J)=D
112        DO 1041 I=1,NX
113        DO 1041 J=1,NY
114        D=0.D0
115        DO 1040 K=1,NU
116  1040  D=D+GM(I,K)*DFM(K,J)
117  1041  DDM(J,I)=D
118        WRITE (6,1014)
119        WRITE (6,1004) ((KM(I,J),I=1,NU),J=1,NX)
120        WRITE (6,1015)
121        WRITE (6,1004) ((FM(I,J),I=1,NX),J=1,NX)
122        WRITE (6,1016)
123        WRITE (6,1004) ((DDM(J,I),I=1,NX),J=1,NY)
124        WRITE (6,1005)
125        WRITE (7,1002) ((KM(I,J),I=1,NU),J=1,NX)
126        WRITE (7,1002) ((FM(I,J),I=1,NX),J=1,NX)
127        WRITE (7,1002) ((DDM(J,I),I=1,NX),J=1,NU)
128        STOP
129        END
```

Table 3-2a
Listing of User-supplied Input Data for Example 3A

```
 1  SYN(MI)   EXAMPLE 3A
 2  MERRIAM
 3  UNIVERSITY OF ROCHESTER
 4            1            3            1
 5   0.                    1.                    0.        -1.
 6  -3.                    1.
 7  -1.                                 3
 8   5.                         0.                 0.    0.
 9   0.                         1.                 0.    0.
10   1.
11   2.
12   1.
```

Table 3-2b
Printed Output from SYN for Example 3A

```
SYN(MI)    EXAMPLE 3A
MERRIAM
UNIVERSITY OF ROCHESTER


NU,NX,NY,ND
  1        3              1              3
FM
   0.000000000D 00    0.100000000D 01    0.000000000D 00    0.000000000D 00
  -0.300000000D 01    0.100000000D 01    0.000000000D 00   -0.100000000D 01
  -0.100000000D 01
GM
   0.500000000D 01    0.000000000D 00    0.000000000D 00
HM
   0.000000000D 00    0.100000000D 01    0.000000000D 00
PM( 1)
   0.100000000D 01
PM( 2)
   0.200000000D 01
PM( 3)
   0.100000000D 01
```

(continued)

Table 3-2b Continued

```
EM(  1)
   0.000000000D 00
EM(  0)
   0.500000000D 01
DM
  -0.100000000D 01   0.300000000D 01   0.200000000D 01
KM
  -0.200000000D 00   0.600000000D 00   0.400000000D 00
AM
   0.100000000D 01   0.100000000D 01  -0.000000000D 00  -0.300000000D 01
  -0.300000000D 01   0.100000000D 01  -0.200000000D 01  -0.100000000D 01
  -0.100000000D 01
BM
   0.100000000D 01   0.000000000D 00   0.000000000D 00
```

where

$$p(s) = \sum_{k=0}^{n} \delta_k \, s^{(n-k)}. \qquad (3.12)$$

Coefficients which appear in these relationships are computed from the well-known relationships [7]

$$\delta_0 = 1, \boldsymbol{\Delta}_0 = \mathbf{I}, \qquad (3.13)$$

$$\delta_k = -\frac{1}{k} \operatorname{tr}(\boldsymbol{\Delta}_{k-1}\mathbf{A}), \; \boldsymbol{\Delta}_k = (\boldsymbol{\Delta}_{k-1}\mathbf{A} + \delta_k \mathbf{I}) \text{ for } k = 1, 2, \ldots, n. \qquad (3.14)$$

Numerator and denominator polynomials of the matrix of transfer functions are factored by technique described in chapter 6 in conjunction with FACTOR and related suprograms.

EVL

Evaluation of the matrix of transfer functions is implemented by program EVL which is summarized as follows:

EVL function: Compute

$$C(s\mathbf{I} - \mathbf{A})^{-1}\mathbf{B}$$

and factor numerator and denominator polynomials.

EVL dummy argument list: None.

EVL input data list: C1/C2/C3/NU,NX,NY,MODE/AM/BM/CM, where $1 \leqslant NU \leqslant 3, 1 \leqslant NX \leqslant 7, 1 \leqslant NY \leqslant 7$.
MODE = 1 for factored form of polynomials or 2 otherwise.
AM(NX,NX) = **A**.
BM(NX,NU) = **B**.
CM(NY,NX) = **C**.

EVL output data list: None.

Subprograms called by EVL: CONVEX,EVAL,FACT,FACTOR,INTL,MATCH, PC,PD,SEARCH,ZEROS.

Program EVL is listed in Table 3-3. There are no user-supplied programs required for program EVL.

Example 3B

Example 3A also serves as an elementary example of the use of program EVL. Matrices **K**, **A**, and **B** which are output data from program SYN are input data for program EVL. A listing of program input cards for this example is given in Table 3-4a, and program printed output is listed in Table 3-4b for this example.

Table 3-3
Listing of EVL

```
 1   C        EVL
 2   C        REAL*8 AM(NX,NX),BM(NX,NU),CM(NY,NX),RM(2,NX),XM(NX,NU),
 3   C       1YM(NX,NX),ZM(NX,NX),ZA(NP,NY,NU),PV(NP),QV(NP),RV(NP),
 4   C       2SV(NP),TV(NP),DV(NP),D,E,DABS
 5   C        INTEGER*4 MV(NX)
 6            REAL*8 AM(7,7),BM(7,3),CM(7,7),RM(2,7),XM(7,3),YM(7,7),
 7           1ZM(7,7),ZA(8,7,3),PV(8),QV(8),RV(8),SV(8),TV(8),DV(8)
 8            REAL*8 D,E,DABS
 9            INTEGER*4 MV(7)
10            LOGICAL*1 COM(65),PRINT
11   1000     FORMAT(65A1)
12   1001     FORMAT(4X,65A1)
13   1002     FORMAT(4D18.10)
14   1003     FORMAT(4I10)
15   1004     FORMAT(4X,4D18.10)
16   1005     FORMAT(4X,2D18.10,I18)
17   1006     FORMAT(' 1')
18   1007     FORMAT(//,'           NU,NX,NY,MODE')
19   1008     FORMAT('          AM')
20   1009     FORMAT('          BM')
21   1010     FORMAT('          CM')
22   1011     FORMAT(//,'           CHAR. POLY.')
23   1012     FORMAT(//,'           TM(',I2,',',I2,')')
24   1013     FORMAT('           NUM. POLY.')
25   1014     FORMAT('           DEN. POLY.')
```

(continued)

Table 3-3 Continued

```
26  1015 FORMAT('          REAL PART,IMAG. PART,MULT.')
27       WRITE (6,1006)
28       DO 1016 I=1,3
29       READ (5,1000) COM
30  1016 WRITE (6,1001) COM
31       READ (5,1003) NU,NX,NY,MODE
32       READ (5,1002) ((AM(I,J),I=1,NX),J=1,NX)
33       READ (5,1002) ((BM(I,J),I=1,NX),J=1,NU)
34       READ (5,1002) ((CM(I,J),I=1,NY),J=1,NX)
35       WRITE (6,1007)
36       WRITE (6,1003) NU,NX,NY,MODE
37       WRITE (6,1008)
38       WRITE (6,1004) ((AM(I,J),I=1,NX),J=1,NX)
39       WRITE (6,1009)
40       WRITE (6,1004) ((BM(I,J),I=1,NX),J=1,NU)
41       WRITE (6,1010)
42       WRITE (6,1004) ((CM(I,J),I=1,NY),J=1,NX)
43       NP=NX+1
44       E=1.D0
45       PRINT=.FALSE.
46       DO 1018 I=1,NX
47       DO 1017 J=1,NX
48  1017 ZM(I,J)=0.D0
49  1018 ZM(I,I)=1.D0
50       DO 1027 L=1,NP
51       PV(L)=E
```

```
      E=0.D0
      DO 1023 I=1,NX
      DO 1020 J=1,NU
      D=0.D0
      DO 1019 K=1,NX
 1019 D=D+ZM(I,K)*BM(K,J)
 1020 XM(I,J)=D
      DO 1022 J=1,NX
      D=0.D0
      DO 1021 K=1,NX
 1021 D=D+ZM(I,K)*AM(K,J)
 1022 YM(I,J)=D
 1023 E=E+YM(I,I)
      E=-E/L
      DO 1025 I=1,NY
      DO 1025 J=1,NU
      D=0.D0
      DO 1024 K=1,NX
 1024 D=D+CM(I,K)*XM(K,J)
 1025 ZA(L,I,J)=D
      DO 1027 I=1,NX
      DO 1026 J=1,NX
 1026 ZM(I,J)=YM(I,J)
 1027 ZM(I,I)=ZM(I,I)+E
      WRITE (6,1011)
      WRITE (6,1004) (PV(I),I=1,NP)
      DO 1038 K=1,NY
```

(continued)

Table 3-3 Continued

```
79          DO 1038 L=1,NU
80          D=DABS(ZA(NP,K,L))
81          DO 1028 I=1,NX
82          M=I
83          IF (D .LT. DABS(ZA(I,K,L))) GO TO 1029
84   1028   CONTINUE
85   1029   NQ=NP-M
86          D=ZA(M,K,L)
87          E=1.D0/D
88          M=M-1
89          DO 1030 I=1,NQ
90   1030   QV(I)=ZA(I+M,K,L)*E
91          CALL PC(NP,PV,NQ,QV,3.D-08,NS,SV)
92          NT=NP
93          DO 1031 I=1,NP
94   1031   TV(I)=PV(I)
95          IF (NS .EQ. 1) GO TO 1033
96          E=1.D0/SV(1)
97          DO 1032 I=1,NS
98   1032   SV(I)=SV(I)*E
99          CALL PD(NQ,QV,NS,SV,3.D-08,NQ,QV,ND,DV)
100         CALL PD(NT,TV,NS,SV,3.D-08,NT,TV,ND,DV)
101  1033   DO 1034 I=1,NQ
102  1034   RV(I)=QV(I)*D
103         WRITE (6,1012) K,L
104         WRITE (6,1013)
```

```
105       WRITE (6,1004) (RV(I),I=1,NQ)
106       IF (MODE .EQ. 2) GO TO 1036
107       IF (NQ .LE. 1) GO TO 1036
108       CALL FACTOR (NQ,QV,1.D0,1.D0,3.D-08,PRINT,J,RM,MV)
109       WRITE (6,1015)
110       DO 1035 I=1,J
111  1035 WRITE (6,1004) RM(1,I),RM(2,I),MV(I)
112  1036 WRITE (6,1014)
113       WRITE (6,1004) (TV(I),I=1,NT)
114       IF (MODE .EQ. 2) GO TO 1038
115       CALL FACTOR(NT,TV,1.D0,1.D0,3.D-08,PRINT,J,RM,MV)
116       WRITE (6,1015)
117       DO 1037 I=1,J
118  1037 WRITE (6,1005) RM(1,I),RM(2,I),MV(I)
119  1038 CONTINUE
120       WRITE (6,1006)
121       STOP
122       END
```

Table 3-4a:
Listing of User-supplied Input Data for Example 3B

```
1   EVL(CONVEX,EVAL,FACT,FACTOR,INTL,MATCH,PC,PD,SEARCH,ZEROS)
2   EXAMPLE 3B  MERRIAM
3   UNIVERSITY OF ROCHESTER
4         1           3           1           1
5    1.         1.         1.
6   -3.         0.                           -3.
7   -1.         -2.
8    1.         0.         0.
9    0.         1.                           -1.
```

Table 3-4b
Printed Output from EVL for Example 3B

```
EVL(CONVEX,EVAL,FACT,FACTOR,INTL,MATCH,PC,PD,SEARCH,ZEROS)
EXAMPLE 3B  MERRIAM
UNIVERSITY OF ROCHESTER

NU,NX,NY,MODE
  1    3    1    1

AM
  0.1000000000D 01   0.1000000000D 01   0.0000000000D 00  -0.3000000000D 01
 -0.3000000000D 01   0.1000000000D 01  -0.2000000000D 01  -0.1000000000D 01
 -0.1000000000D 01
```

```
BM
   0.1000000000D 01    0.0000000000D 00    0.0000000000D 00
CM
   0.0000000000D 00    0.1000000000D 01    0.0000000000D 00

CHAR. POLY.
   0.1000000000D 01    0.3000000000D 01    0.3000000000D 01    0.1000000000D 01

TM( 1,  1)
 NUM. POLY.
   0.1000000000D 01
 DEN. POLY.
   0.1000000000D 01    0.2000000000D 01    0.1000000000D 01
 REAL PART, IMAG. PART, MULT.
  -0.1000000000D 01    0.0000000000D 00    2
```

4

Optimization with Functionals Defined on an Infinite Time Interval

Computer programs presented in this chapter are intended to be representative of numerous examples of parameter optimization applied to design problems which are defined on an infinite time interval. Chapters 4 and 5 of *Automated Design of Control Systems* [1] contain numerous additional examples of such design problems that could have also been included here.

4.1 Optimal Gain Control

Optimal gain control of linear systems with deterministic signals is an important class of design problems which can be solved readily by the set of computer programs discussed in this section.

Description of the Basic Method

The linear system used for optimal gain control is defined by

$$\dot{x} = Fx + Gu, x(0) = x_o , \quad (4.1)$$

$$y = Hx , \quad (4.2)$$

and

$$e = Dx + Eu . \quad (4.3)$$

Coefficient matrices **D** and **E** may of course be specified by the excess-pole-specification method given in chapter 3. The performance integral used for optimal gain is taken to be

$$I = \int_0^\infty \left\{ \frac{1}{2} x'Qx + \frac{1}{2} e'e \right\} dt , \quad (4.4)$$

where **Q** is a positive semidefinite matrix which is often taken to be **0**.

The simplest special case of optimal gain control corresponds to a control equation

$$u = -Cx \qquad (4.5)$$

which includes the entire state vector. Parameters, which are selected via optimization techniques, are elements of the feedback gain matrix **C**. Optimal gains are given by

$$C = (E'E)^{-1} (E'D + G'X) \qquad (4.6)$$

where

$$0 = P + XA + A'X - XRX, \qquad (4.7)$$

and coefficient matrices **P**, **A**, and **R** are suitably specified in terms coefficient matrices **D**, **E**, **F**, **G**, and **Q**.

The first three computer programs described in this section facilitate the solution of linear optimal control problems. Program FORM1 is used to compute coefficient matrices **P**, **A**, and **R**. Program QUAD can then be used to solve the algebraic Riccati equations given in equation (4.7) by the Newton method of successive substitutions that forms the constructive proof of Theorem 5-6 appearing in *Automated Design of Control Systems* [1]. Program GAIN1 can finally be used to compute the optimal feedback gain matrix that is given in equation (4.6).

The more general case of optimal gain control corresponds to a control equation

$$u = -Cy \qquad (4.8)$$

which may not include the entire state vector, depending on how coefficient matrix **H** is defined. Subprogram LOAD1 is used to specialize subprograms MIN1 or MIN3 so that methods of successive approximation can be employed to find optimal feedback gains. Subprogram LOAD1 computes objective function

$$f = tr\{T'XT\}, \qquad (4.9)$$

where

$$0 = W + XA + A'X, \qquad (4.10)$$

and its gradient f_C. Algebraic Lyapunov equations given in equation (4.10) are solved by subprogram LIN1 by methods given in chapter 6. Coefficient matrices **W** and **A** are suitably specified in terms of coefficient matrices **D**, **E**, **F**, **G**, **H** and feedback gain matrix **C**.

FORM1

Computation of coefficient matrices **P**, **A**, and **R** is performed by FORM1, which is summarized as follows:

FORM1 function: Compute

$$A = F - G(E'E)^{-1}E'D ,$$
$$P = D'[I - E(E'E)^{-1}E']D + Q ,$$

and

$$R = G(E'E)^{-1}G'$$

for use with QUAD.

FORM1 dummy argument list: None.

FORM1 Input Data List: C1/C2/C3/NU,NX/FM/GM/DM/EM/QM, where $1 \leqslant NU \leqslant 3, 1 \leqslant NX \leqslant 7$.
FM(NX,NX) = **F**.
GM(NX,NU) = **G**.
DM(NU,NX) = **D**.
EM(NU,NU) = **E**.
QM(NX,NX) = **Q**.

FORM1 output data list: AM/PM/RM, where
AM(NX,NX) = **A**.
PM(NX,NX) = **P**.
RM(NX,NX) = **R**.

Subprograms called by FORM1: MI.

Program FORM1 is listed in Table 4-1. User problems are specified entirely by input data so that a user-supplied subprogram is not needed for the use of FORM1.

Table 4-1
Listing of FORM1

```
1    C      FORM1
2    C      REAL*8 AM(NX,NX),DM(NU,NX),EM(NU,NU),FM(NX,NX),GM(NX,NU),
3    C     1PM(NX,NX),QM(NX,NX),RM(NX,NX),DDM(NU,NX),DEM(NU,NU),
4    C     2DFM(NX,NU),D
5           REAL*8 AM(7,7),DM(3,7),EM(3,3),FM(7,7),GM(7,3),PM(7,7),
6          1QM(7,7),RM(7,7),DDM(3,7),DEM(3,3),DFM(7,3)
7           DATA MU/3/
8           REAL*8 D
9           LOGICAL*1 COM(65)
10   1000   FORMAT(65A1)
11   1001   FORMAT(4X,65A1)
12   1002   FORMAT(4D18.10)
13   1003   FORMAT(2I10)
14   1004   FORMAT(4X,4D18.10)
15   1005   FORMAT(' 1')
16   1006   FORMAT(//,'             NU,NX')
17   1007   FORMAT('           FM')
18   1008   FORMAT('           GM')
19   1009   FORMAT('           DM')
20   1010   FORMAT('           EM')
21   1011   FORMAT('           QM')
22   1012   FORMAT(//,'           AM')
23   1013   FORMAT('           PM')
24   1014   FORMAT('           RM')
```

```
25        WRITE (6,1005)
26        DO 1015 I=1,3
27        READ (5,1000) COM
28  1015  WRITE (6,1001) COM
29        READ (5,1003) M,N
30        READ (5,1002) ((FM(I,J),I=1,N),J=1,N)
31        READ (5,1002) ((GM(I,J),I=1,N),J=1,M)
32        READ (5,1002) ((DM(I,J),I=1,M),J=1,M)
33        READ (5,1002) ((EM(I,J),I=1,M),J=1,M)
34        READ (5,1002) ((QM(I,J),I=1,J),J=1,N)
35        WRITE (6,1006)
36        WRITE (6,1003) M,N
37        WRITE (6,1007)
38        WRITE (6,1004) ((FM(I,J),I=1,N),J=1,N)
39        WRITE (6,1008)
40        WRITE (6,1004) ((GM(I,J),I=1,N),J=1,M)
41        WRITE (6,1009)
42        WRITE (6,1004) ((DM(I,J),I=1,M),J=1,M)
43        WRITE (6,1010)
44        WRITE (6,1004) ((EM(I,J),I=1,M),J=1,M)
45        WRITE (6,1011)
46        WRITE (6,1004) ((QM(I,J),I=1,J),J=1,N)
47        DO 1017 I=1,M
48        DO 1017 J=I,M
49        D=0.D0
50        DO 1016 K=1,M
51  1016  D=D+EM(K,I)*EM(K,J)
52        DEM(I,J)=D
```

(continued)

Table 4-1 Continued

```
53      1017  DEM(J,I)=D
54            CALL MI(MU,M,DEM)
55            DO 1019 I=1,M
56            DO 1019 J=1,N
57            D=0.D0
58            DO 1018 K=1,M
59      1018  D=D+EM(K,I)*DM(K,J)
60      1019  DDM(I,J)=D
61            DO 1021 I=1,N
62            DO 1021 J=1,M
63            D=0.D0
64            DO 1020 K=1,M
65      1020  D=D+GM(I,K)*DEM(K,J)
66      1021  DFM(I,J)=D
67            DO 1025 I=1,N
68            DO 1023 J=1,N
69            D=FM(I,J)
70            DO 1022 K=1,M
71      1022  D=D-DFM(I,K)*DDM(K,J)
72      1023  AM(I,J)=D
73            DO 1025 J=I,N
74            D=0.D0
75            DO 1024 K=1,M
76      1024  D=D+DFM(I,K)*GM(J,K)
77      1025  RM(I,J)=D
```

OPTIMIZATION ON AN INFINITE TIME INTERVAL

```
78          DO 1027 I=1,M
79          DO 1027 J=1,N
80          D=0.D0
81          DO 1026 K=1,M
82 1026     D=D+DEM(I,K)*DDM(K,J)
83 1027     DFM(J,I)=D
84          DO 1029 I=1,N
85          DO 1029 J=I,N
86          D=QM(I,J)
87          DO 1028 K=1,M
88 1028     D=D+DM(K,I)*DM(K,J)-DDM(K,I)*DFM(J,K)
89 1029     PM(I,J)=D
90          WRITE (6,1012)
91          WRITE (6,1004) ((AM(I,J),I=1,N),J=1,N)
92          WRITE (6,1013)
93          WRITE (6,1004) ((PM(I,J),I=1,N),J=1,N)
94          WRITE (6,1014)
95          WRITE (6,1004) ((RM(I,J),I=1,N),J=1,N)
96          WRITE (6,1005)
97          WRITE (7,1002) ((AM(I,J),I=1,N),J=1,N)
98          WRITE (7,1002) ((PM(I,J),I=1,N),J=1,N)
99          WRITE (7,1002) ((RM(I,J),I=1,N),J=1,N)
100         STOP
101         END
```

Example 4A

An example of the use of program FORM1 for the computation of matrices **P**, **A**, and **R** is the optimal linear control problem corresponding to

$$\mathbf{F} = \begin{bmatrix} 0 & 0 \\ 1 & 0 \end{bmatrix}, \mathbf{G} = \begin{bmatrix} 1 \\ 0 \end{bmatrix}, \mathbf{D} = [2\ 1], \mathbf{E} = [1]$$

and

$$\mathbf{Q} = \begin{bmatrix} 0 & 0 \\ 0 & 0 \end{bmatrix}.$$

A listing of program input cards for this example is given in Table 4-2a, and corresponding program printed output is listed in Table 4-2b.

QUAD

Computation of the solution to algebraic Riccati equations can be accomplished using program QUAD with NV = 0. The case with NV > 0 is explained in Section 4.3, "Simulation of Human Controller Models Based on Optimization." Program QUAD is summarized as follows:

QUAD function: Compute the solution to

$$\mathbf{XA} + \mathbf{A'X} + \sum_{k=1}^{m} \mathbf{B}'_k \mathbf{XB}_k + \mathbf{P} - \mathbf{XRX} = 0.$$

QUAD dummy argument list: None.

QUAD input data list: C1/C2/C3/NI,MODE,NX,NV/DXMIN/AM/BA/PM/RM/ XM, where
  NI = maximum number of iterations.
  MODE = 1 for use with the initialization step to obtain $\dot{\mathbf{X}}$ negative semidefinite or 2 otherwise.
  $1 \leqslant NX \leqslant 7$, $0 \leqslant NV \leqslant 3$.
  DXMIN = minimum value of $\|\dot{\mathbf{X}}\|$ requiring additional iterations.
  AM(NX,NX) = **A**.
  BA(NV,NX,NX) = $\mathbf{B}_1/\mathbf{B}_2/\ldots$.
  PM(NX,NX) = **P**.
  RM(NX,NX) = **R**.
  XM(NX,NX) = positive semidefinite $\mathbf{X}_o$ for which the equivalent coefficient matrix of

Table 4-2a
Listing of User-supplied Input Data for Example 4A

```
1   FORM1(MI)   EXAMPLE 4A
2   MERRIAM
3   UNIVERSITY OF ROCHESTER
4              1          2                    0.
5       0.                1.
6       1.                0.         0.
7       2.                1.
8       1.
9       0.                          0.
```

Table 4-2b
Printed Output from FORM1 for Example 4A

```
FORM1(MI)    EXAMPLE 4A
MERRIAM
UNIVERSITY OF ROCHESTER

 NU, NX
    1        2

 FM
    0.0000000000D 00   0.1000000000D 01   0.0000000000D 00   0.0000000000D 00
 GM
    0.1000000000D 01   0.0000000000D 00
 DM
    0.2000000000D 01   0.1000000000D 01
 EM
    0.1000000000D 01
 QM
    0.0000000000D 00   0.0000000000D 00   0.0000000000D 00
 AM
   -0.2000000000D 01   0.1000000000D 01  -0.1000000000D 01  -0.0000000000D 00
 PM
    0.0000000000D 00   0.0000000000D 00   0.0000000000D 00
 RM
    0.1000000000D 01   0.0000000000D 00   0.0000000000D 00
```

OPTIMIZATION ON AN INFINITE TIME INTERVAL　　　　　101

$$X(A - RX_o) + (A - RX_o)'X + \sum_{k=1}^{m} B'_k X B_k$$

is asymptotically stable.

QUAD output data list: XM, where
　　XM(NX,NX) = X.

Subprograms called by QUAD: DXM,INC,LIN2,MI.

Program QUAD is listed in Table 4-3. User problems again are specified entirely by input data so that a user-supplied subprogram is not needed for the use of QUAD.

Example 4B

An example of the use of program QUAD for the solution of equation (4.7) is the linear optimal control problem specified in Example 4A. A listing of program input cards for this example is given in Table 4-4a, and corresponding printed output is listed in Table 4-4b. Quadratic convergence is illustrated by these data. In this particular example, condition $X = 0$ occurs, which corresponds to condition $I = 0$ for all x_o and hence to condition $e(t) = 0$.

GAIN1

Computation of the feedback gain matrix for linear optimal control can now be accomplished using program GAIN1 which is summarized as follows:

GAIN1 function: Compute

$$C = (E'E)^{-1}(E'D + G'X).$$

GAIN1 dummy argument list: None.

GAIN1 input data list: C1/C2/C3/NU,NX/DM/EM/GM/XM, where
　　$1 \leqslant NU \leqslant 3, 1 \leqslant NX \leqslant 7$.
　　DM(NU,NX) = D.
　　EM(NU,NU) = E.
　　GM(NX,NU) = G.
　　XM(NX,NX) = X.

GAIN1 output data list: CM, where
　　CM(NU,NX) = C.

Table 4-3
Listing of QUAD

```
1    C     QUAD
2    C     REAL*8 AM(NX,NX),BA(NV,NX,NX),PM(NX,NX),RM(NX,NX),
3    C    1XM(NX,NX),YM(NX,NX),ZM(NX,NX),CM(NT,NT),DXMIN,D,DABS
4          REAL*8 AM(7,7),BA(3,7,7),PM(7,7),RM(7,7),
5         1XM(7,7),YM(7,7),ZM(7,7),CM(28,28)
6          DATA MT/28/,MX/7/,MV/3/
7          REAL*8 DXMIN,D,DABS
8          LOGICAL*1 COM(65)
9    1000  FORMAT(65A1)
10   1001  FORMAT(4X,65A1)
11   1002  FORMAT(4D18.10)
12   1003  FORMAT(4I10)
13   1004  FORMAT(4X,4D18.10)
14   1005  FORMAT('1')
15   1006  FORMAT(//,'        NI,MODE,NX,NV')
16   1007  FORMAT('        AM')
17   1008  FORMAT('        DXMIN')
18   1009  FORMAT('        BM(',I2,')')
19   1010  FORMAT('        PM')
20   1011  FORMAT('        RM')
21   1012  FORMAT('        XM')
22   1013  FORMAT(//,'        ITERATION=',I2,'   ||DXM||=',D18.10)
23   1014  FORMAT(//,'        DESIGN COMPLETE')
```

OPTIMIZATION ON AN INFINITE TIME INTERVAL

(continued)

```
24  1015  FORMAT(//,'           TRIAL LIMIT')
25        WRITE (6,1005)
26        DO 1016 I=1,3
27        READ (5,1000) COM
28  1016  WRITE (6,1001) COM
29        READ (5,1003) NI,MODE,NX,NV
30        READ (5,1002) DXMIN
31        READ (5,1002) ((AM(I,J),I=1,NX),J=1,NX)
32        IF (NV .EQ. 0) GO TO 1018
33        DO 1017 K=1,NV
34  1017  READ (5,1002) ((BA(K,I,J),I=1,NX),J=1,NX)
35  1018  READ (5,1002) ((PM(I,J),I=1,J),J=1,NX)
36        READ (5,1002) ((RM(I,J),I=1,J),J=1,NX)
37        READ (5,1002) ((XM(I,J),I=1,J),J=1,NX)
38        WRITE (6,1006)
39        WRITE (6,1003) NI,MODE,NX,NV
40        WRITE (6,1008)
41        WRITE (6,1004) DXMIN
42        WRITE (6,1007)
43        WRITE (6,1004) ((AM(I,J),I=1,NX),J=1,NX)
44        IF (NV) 1021,1021,1019
45  1019  DO 1020 K=1,NV
46        WRITE (6,1009) K
47  1020  WRITE (6,1004) ((BA(K,I,J),I=1,NX),J=1,NX)
48  1021  WRITE (6,1010)
49        WRITE (6,1004) ((PM(I,J),I=1,J),J=1,NX)
50        WRITE (6,1011)
51        WRITE (6,1004) ((RM(I,J),I=1,J),J=1,NX)
52        WRITE (6,1012)
```

Table 4-3 Continued

```
53              WRITE (6,1004) ((XM(I,J),I=1,J),J=1,NX)
54              DO 1022 I=1,NX
55              DO 1022 J=I,NX
56              RM(J,I)=RM(I,J)
57         1022 XM(J,I)=XM(I,J)
58              IF (MODE .EQ. 2) GO TO 1029
59              CALL DXM(MX,MV,NX,NV,AM,BA,PM,RM,XM,YM)
60              DO 1025 I=1,NX
61              D=-2.D0*YM(I,I)
62              DO 1023 J=1,NX
63              D=D+DABS(YM(I,J))
64         1023 ZM(I,J)=0.D0
65              IF (D) 1024,1025,1025
66         1024 D=0.D0
67         1025 ZM(I,I)=D
68              DO 1027 I=1,NX
69              DO 1027 J=1,NX
70              D=AM(I,J)
71              DO 1026 K=1,NX
72         1026 D=D-RM(I,K)*XM(K,J)
73         1027 YM(I,J)=D
74              CALL INC(MT,MX,MV,NX,NV,YM,BA,CM)
75              CALL LIN2(MT,MX,NX,CM,ZM,YM)
76              DO 1028 I=1,NX
77              DO 1028 J=1,NX
78         1028 XM(I,J)=XM(I,J)+YM(I,J)
79         1029 DO 1034 L=1,NI
80              CALL DXM(MX,MV,NX,NV,AM,BA,PM,RM,XM,YM)
```

```
81            D=0.D0
82            DO 1030 I=1,NX
83            DO 1030 J=I,NX
84            IF (D .GE. DABS(YM(I,J))) GO TO 1030
85            L=DABS(YM(I,J))
86       1030 CONTINUE
87            WRITE (6,1013) L,D
88            WRITE (6,1012)
89            WRITE (6,1004) ((XM(I,J),I=1,J),J=1,NX)
90            IF (D .GE. DXMIN) GO TO 1031
91            WRITE (6,1014)
92            GO TO 1035
93       1031 DO 1033 I=1,NX
94            DO 1033 J=1,NX
95            D=AM(I,J)
96            DO 1032 K=1,NX
97       1032 D=D-RM(I,K)*XM(K,J)
98       1033 ZM(I,J)=D
99            CALL INC(MT,MX,MV,NX,NV,ZM,BA,CM)
100           CALL LIN2(MT,MX,NX,CM,YM,ZM)
101           DO 1034 I=1,NX
102           DO 1034 J=1,NX
103      1034 XM(I,J)=XM(I,J)+ZM(I,J)
104           WRITE (6,1015)
105      1035 WRITE (6,1005)
106           WRITE (7,1002) ((XM(I,J),I=1,J),J=1,NX)
107           STOP
108           END
```

Table 4-4a
Listing of User-supplied Input Data for Example 4B

```
1   QUAD(DXM,INC,LIN2,MI)   EXAMPLE 4B
2   MERRIAM
3   UNIVERSITY OF ROCHESTER
4           7           1           2           0
5    0.
6   -2.         1.          0.         -1.
7    0.         0.          2.          0.
8    1.         0.                      0.
9    2.         1.                      2.
```

Table 4–4b
Printed Output from QUAD for Example 4B

```
QUAD(DXM,INC,LIN2,MI)    EXAMPLE 4B
MERRIAM
UNIVERSITY OF ROCHESTER


NI,MODE,NX,NV
   7      1        2         0
DXMIN
  0.000000000D 00
AM
 -0.200000000D 01   0.100000000D 01  -0.100000000D 01   0.000000000D 00
PM
  0.000000000D 00   0.000000000D 00   0.000000000D 00
RM
  0.100000000D 01   0.000000000D 00   0.000000000D 00
XM
  0.200000000D 01   0.100000000D 01   0.200000000D 01

ITERATION= 1    ||DXM||= 0.696289062D 02
XM
  0.706250000D 01   0.425000000D 01   0.251250000D 02
```

(continued)

Table 4-4b Continued

ITERATION= 2 ||DXM||=   0.16980506130 02
XM
  0.29417590310 01   0.17202380950 01   0.10182676520 01

ITERATION= 3 ||DXM||=   0.38263191610 01
XM
  0.98566108860 00   0.54392648740 00   0.30866052130 00

ITERATION= 4 ||DXM||=   0.62547717040 00
XM
  0.19478994210 00   0.95812859660-01   0.50678904180-01

ITERATION= 5 ||DXM||=   0.33943475550-01
XM
  0.10552390100-01   0.41887189020-02   0.20934215110-02

ITERATION= 6 ||DXM||= 0.110677824 3D-03
XM
  0.320372444 7D-04 0.873608999 5D-05 0.553481021 4D-05

ITERATION= 7 ||DXM||= 0.102636737 0D-08
XM
  0.275671492 9D-09 0.381593008 4D-10 0.721134744 8D-10

TRIAL LIMIT

Subprograms called by GAIN1: MI.

Program GAIN1 is listed in Table 4-5. No user-supplied subprograms are required for the use of GAIN1.

Example 4C

Program GAIN1 is illustrated using the linear optimal control problem specified in Example 4A. Program input cards are listed in Table 4-6a, and corresponding program printed output from GAIN1 is listed in Table 4-6b.

LOAD1

Subprogram LOAD1 is used to specialize either MIN1 or MIN3 for solving the optimal gain control problem without the necessity of computing second partial derivatives with respect to feedback gains. This subprogram is used to compute objective function $f(\mathbf{C})$ and its gradient $f_\mathbf{C}$, and this subprogram is summarized as follows:

LOAD1 function: Compute f and $f_\mathbf{C}$ for optimal gain control problems defined by
  $\mathbf{A} = \mathbf{F} - \mathbf{GCH}, \mathbf{W} = (\mathbf{D} - \mathbf{ECH})' (\mathbf{D} - \mathbf{ECH})$,
  and $\mathbf{f} = tr\left\{\mathbf{T'XT}\right\}$
as a subprogram for use with MIN1S or MIN3S.

LOAD1 dummy argument list: MC,CV,F,DFV,MODE, where
  $1 \leqslant MC \leqslant 21$.
  CV(MC) = c.
  F = f
  DFV(MC) = f_c
  MODE = 1 for initialization and 2 otherwise.

LOAD1 input data list: NU,NX,NF,NP/FM/GM/HM/DM/EM/TM, where
  $1 \leqslant NU \leqslant 3, 1 \leqslant NX \leqslant 7, 1 \leqslant NF \leqslant 7, 1 \leqslant NP \leqslant 7$.
  FM(NX,NX) = F.
  GM(NX,NU) = G.
  HM(NF,NX) = H.
  DM(NU,NX) = D.
  EM(NU,NU) = E.
  TM(NX,NP) = T.

LOAD1 output data list: None.

Subprograms called by LOAD1: LIN1,MI.

Table 4-5
Listing of GAIN1

```
1     C     GAIN1
2     C     REAL*8  DM(NU,NX),EM(NU,NU),GM(NX,NU),XM(NX,NX),DDM(NU,NX),
3     C    1DEM(NU,NU),D
4           REAL*8 DM(3,7),EM(3,3),GM(7,3),XM(7,7),DDM(3,7),DEM(3,3)
5           DATA MU/3/
6           REAL*8 D
7           LOGICAL*1 COM(65)
8     1000  FORMAT(65A1)
9     1001  FORMAT(4X,65A1)
10    1002  FORMAT(4D18.10)
11    1003  FORMAT(2I10)
12    1004  FORMAT(4X,4D18.10)
13    1005  FORMAT('1')
14    1006  FORMAT('       DM')
15    1007  FORMAT('       EM')
16    1008  FORMAT('       GM')
17    1009  FORMAT('       XM')
18    1010  FORMAT(//,'       CM')
19    1011  FORMAT(//,'       NU,NX')
20          WRITE (6,1005)
21          DO 1012 I=1,3
22          READ (5,1000) COM
23          WRITE (6,1001) COM
24    1012  READ (5,1003) M,N
```

Table 4-5 Continued

```
25        READ  (5,1002)   ((DM(I,J),I=1,M),J=1,N)
26        READ  (5,1002)   ((EM(I,J),I=1,M),J=1,M)
27        READ  (5,1002)   ((GM(I,J),I=1,N),J=1,M)
28        READ  (5,1002)   ((XM(I,J),I=1,J),J=1,N)
29        WRITE (6,1011)
30        WRITE (6,1003)   M,N
31        WRITE (6,1006)
32        WRITE (6,1004)   ((DM(I,J),I=1,M),J=1,N)
33        WRITE (6,1007)
34        WRITE (6,1004)   ((EM(I,J),I=1,M),J=1,M)
35        WRITE (6,1008)
36        WRITE (6,1004)   ((GM(I,J),I=1,N),J=1,M)
37        WRITE (6,1009)
38        WRITE (6,1004)   ((XM(I,J),I=1,J),J=1,N)
39        DO 1013 I=1,N
40        DO 1013 J=I,N
41  1013  XM(J,I)=XM(I,J)
42        DO 1015 I=1,M
43        DO 1015 J=I,M
44        D=0.D0
45        DO 1014 K=1,M
46  1014  D=D+EM(K,I)*EM(K,J)
47        DEM(I,J)=D
48  1015  DEM(J,I)=D
49        CALL MI(MU,M,DEM)
```

```
50          DO 1018 I=1,M
51          DO 1018 J=1,N
52          D=0.D0
53          DO 1016 K=1,M
54 1016     D=D+EM(K,I)*DM(K,J)
55          DO 1017 K=1,N
56 1017     D=D+GM(K,I)*XM(K,J)
57 1018     DDM(I,J)=D
58          DO 1020 I=1,M
59          DO 1020 J=1,N
60          D=0.D0
61          DO 1019 K=1,M
62 1019     D=D+DEM(I,K)*DDM(K,J)
63 1020     DM(I,J)=D
64          WRITE (6,1010)
65          WRITE (6,1004) ((DM(I,J),I=1,M),J=1,N)
66          WRITE (6,1005)
67          WRITE (7,1002) ((DM(I,J),I=1,M),J=1,N)
68          STOP
69          END
```

Table 4-6a
Listing of User-supplied Input Data for Example 4C

```
1    GAIN1(MI)    EXAMPLE 4C
2    MERRIAM
3    UNIVERSITY OF ROCHESTER
4              1              2
5          2.             1.
6          1.
7          1.             0.
8            0.             0.              0.
```

Subprogram LOAD1 is listed in Table 4-7. Either MIN1C and MIN1S or MIN3C or MIN3S must be supplied by the user in order to employ subprogram LOAD1 with either subprogram MIN1 or MIN3.

Example 4D

Subprogram LOAD1 is illustrated using the optimal gain control problem specified in Example 4A. Program MIN1C and subprogram MIN1S are supplied by the user and are listed in Table 4-8a. Program input cards, which are also supplied by the user, are listed in Table 4-8b, and corresponding program printed outputs from MIN1 and LOAD1 are listed in Table 4-8c. Output for this example illustrates use of the conjugate-gradient method of a nonquadratic objective function.

4.2 Optimal Linear Filters

Design of linear filters is an application of optimization techniques that arises frequently in the context of control. Moreover, the optimal filter problem is closely associated with the linear optimal control problem.

Description of the Basic Method

Stochastic equations for a target and an observer are taken to be

$$d\mathbf{x} = (\mathbf{Fx})dt + \mathbf{G}d\mathbf{u}, \mathbf{x}(0) = \mathbf{x}_o \qquad (4.11)$$

Table 4-6b
Printed Output from GAIN1 for Example 4C

```
GAIN1(MI)    EXAMPLE 4C
MERRIAM
UNIVERSITY OF ROCHESTER

NU,NX
    1         2
DM
    0.2000000000D 01    0.1000000000D 01
EM
    0.1000000000D 01
GM
    0.1000000000D 01    0.0000000000D 00
XM
    0.0000000000D 00    0.0000000000D 00    0.0000000000D 00

CM
    0.2000000000D 01    0.1000000000D 01
```

Table 4-7
Listing of LOAD1

```
      SUBROUTINE LOAD1(MC,CV,F,DFV,MODE)
C     OPTIMAL GAIN CONTROL SUBPROGRAM FOR MIN1 AND MIN3
      REAL*8 AM(NX,NX),BM(NX,NX),CM(NU,NF),DM(NU,NX),EM(NU,NU),
     1FM(NX,NX),GM(NX,NU),HM(NF,NX),TM(NX,NF),DCM(NU,NX),
     2DDM(NX,NX),WM(NX,NX),XM(NX,NX),YM(NX,NF),CV(NC),DFV(NC),D,F
      REAL*8 CV(MC),DFV(MC),F
      REAL*8 AM(7,7),BM(7,7),CM(3,7),DM(3,7),EM(3,3),FM(7,7),
     1GM(7,3),HM(7,7),TM(7,7),DCM(3,7),DDM(7,7),WM(7,7),
     2XM(7,7),YM(7,7)
      DATA MX/7/
      REAL*8 D
 1000 FORMAT(4D18.10)
 1001 FORMAT(4I10)
 1002 FORMAT(4X,4D18.10)
 1003 FORMAT(//,'        NU,NX,NF,NP')
 1004 FORMAT('       FM ')
 1005 FORMAT('       GM ')
 1006 FORMAT('       HM ')
 1007 FORMAT('       DM ')
 1008 FORMAT('       EM ')
 1009 FORMAT('       TM ')
      GO TO (1010,1011),MODE
 1010 READ (5,1001) M,N,NF,NP
```

```
24      READ (5,1000) ((FM(I,J),I=1,N),J=1,N)
25      READ (5,1000) ((GM(I,J),I=1,N),J=1,M)
26      READ (5,1000) ((HM(I,J),I=1,NF),J=1,N)
27      READ (5,1000) ((DM(I,J),I=1,M),J=1,N)
28      READ (5,1000) ((EM(I,J),I=1,M),J=1,M)
29      READ (5,1000) ((TM(I,J),I=1,N),J=1,NP)
30      WRITE (6,1003)
31      WRITE (6,1001) M,N,NF,NP
32      WRITE (6,1004)
33      WRITE (6,1002) ((FM(I,J),I=1,N),J=1,N)
34      WRITE (6,1005)
35      WRITE (6,1002) ((GM(I,J),I=1,N),J=1,M)
36      WRITE (6,1006)
37      WRITE (6,1002) ((HM(I,J),I=1,NF),J=1,N)
38      WRITE (6,1007)
39      WRITE (6,1002) ((DM(I,J),I=1,M),J=1,N)
40      WRITE (6,1008)
41      WRITE (6,1002) ((EM(I,J),I=1,M),J=1,M)
42      WRITE (6,1009)
43      WRITE (6,1002) ((TM(I,J),I=1,N),J=1,NP)
44      RETURN
45 1011 K=0
46      DO 1012 J=1,NF
47      DO 1012 I=1,M
48      K=K+1
49 1012 CM(I,J)=CV(K)
50      DO 1014 I=1,M
51      DO 1014 J=1,N
52      D=0.D0
```

(continued)

Table 4-7 Continued

```
53            DO 1013  K=1,NF
54     1013   D=D+CM(I,K)*HM(K,J)
55     1014   DCM(I,J)=D
56            DO 1016  I=1,N
57            DO 1016  J=1,N
58            D=FM(I,J)
59            DO 1015  K=1,M
60     1015   D=D-GM(I,K)*DCM(K,J)
61            AM(I,J)=D
62     1016   BM(J,I)=D
63            DO 1018  I=1,M
64            DO 1018  J=1,N
65            D=DM(I,J)
66            DO 1017  K=1,M
67     1017   D=D-EM(I,K)*DCM(K,J)
68     1018   DDM(I,J)=D
69            DO 1020  I=1,N
70            DO 1020  J=1,N
71            D=0.D0
72            DO 1019  K=1,M
73     1019   D=D+DDM(K,I)*DDM(K,J)
74            WM(I,J)=D
75     1020   WM(J,I)=D
76            CALL LIN1(MX,MX,N,N,AM,BM,WM,XM)
77            DO 1022  I=1,N
```

```
78         DO 1022 J=I,N
79         D=0.D0
80         DO 1021 K=1,NP
81   1021  D=D+TM(I,K)*TM(J,K)
82         WM(I,J)=D
83   1022  WM(J,I)=D
84         CALL LIN1(MX,MX,N,N,BM,AM,WM,DDM)
85         DO 1024 I=1,N
86         DO 1024 J=1,NF
87         D=0.D0
88         DO 1023 K=1,N
89   1023  D=D+DDM(I,K)*HM(J,K)
90   1024  YM(I,J)=D
91         DO 1026 I=1,N
92         DO 1026 J=1,NP
93         D=0.D0
94         DO 1025 K=1,N
95   1025  D=D+XM(I,K)*TM(K,J)
96   1026  DDM(I,J)=D
97         F=0.D0
98         DO 1027 I=1,NP
99         DO 1027 K=1,N
100  1027  F=F+TM(K,I)*DDM(K,I)
101        DO 1032 I=1,M
102        DO 1029 J=I,M
103        D=0.D0
104        DO 1028 K=1,M
```

(continued)

Table 4-7 Continued

```
105   1028   D=D+EM(K,I)*EM(K,J)
106          WM(I,J)=D
107   1029   WM(J,I)=D
108          DO 1032 J=1,N
109          D=0.D0
110          DO 1030 K=1,M
111   1030   D=D+EM(K,I)*DM(K,J)
112          DO 1031 K=1,N
113   1031   D=D+GM(K,I)*XM(K,J)
114   1032   DDM(I,J)=D
115          DO 1034 I=1,M
116          DO 1034 J=1,N
117          D=-DDM(I,J)
118          DO 1033 K=1,M
119   1033   D=D+WM(I,K)*DCM(K,J)
120   1034   DDM(I,J)=D
121          II=0
122          DO 1036 I=1,M
123          DO 1036 J=1,NF
124          II=II+1
125          D=0.D0
126          DO 1035 K=1,N
127   1035   D=D+DDM(I,K)*YM(K,J)
128   1036   DFV(II)=2.D0*D
129          RETURN
130          END
```

Table 4-8a
Listing of User-supplied Programs for Example 4D

```
C     MIN1C  EXAMPLE 4D
       CALL MIN1
       STOP
       END
       SUBROUTINE MIN1S(MC,CV,F,DFV,MODE)
       REAL*8 CV(MC),DFV(MC),F
       CALL LOAD1(MC,CV,F,DFV,MODE)
       RETURN
       END
```

Table 4-8b
Listing of User-supplied Input Data for Example 4D

```
 1  MIN1(LIN1,LOAD1,MI,MIN1S)   EXAMPLE 4D
 2  MERRIAM
 3  UNIVERSITY OF ROCHESTER
 4       4              4             2
 5  .01            0.            0.
 6  .1             1.            0.5
 7  2.1            1.
 8                 2             2
 9  0.             1.            1.
10  1.             0.            0.           10.
11  1.             0.            1.           10.
12  .2
13  1.             1.                          0.
14  0.                                         1.
```

Table 4-8c
Printed Output from MIN1 and LOAD1 for Example 4D

```
MIN1(LIN1,LOAD1,MI,MIN1S)       EXAMPLE 4D
MERRIAM
UNIVERSITY OF ROCHESTER


NI,NT,NR,NC
  4    4         2         2
RATIO,DFMIN,FMIN,FMAX
 0.1000000000D-01  0.0000000000D 00  0.0000000000D 00  0.1000000000D 02
ALMIN,ALMAX,RMIN,RMAX
 0.1000000000D 00  0.1000000000D 01  0.5000000000D 00  0.1000000000D 02


NU,NX,NF,NP
  1    2         2         1
FM
 0.0000000000D 00  0.1000000000D 01  0.0000000000D 00
GM
 0.1000000000D 01  0.0000000000D 00
HM
 0.1000000000D 01  0.0000000000D 00  0.0000000000D 00  0.1000000000D 01
DM
 0.2000000000D 01  0.1000000000D 01
EM
 0.1000000000D 01
```

(continued)

Table 4-8c Continued

TM
   0.000000000D 00   0.100000000D 01

ITERATION= 0   TRIAL= 0   F= 0.238095238 1D-02
ALPHA= 0.000000000D 00   DF= -0.116903579 3D-01 |DFV| =   0.108121958 6D 00
CV
   0.210000000D 01   0.100000000D 01

ITERATION= 0   TRIAL= 1   F= 0.380437398 8D-03
ALPHA= 0.407336096 3D 00   DF= 0.170195386 3D-02 |DFV| =   0.176937921 0D-01
CV
   0.208106487 5D 01   0.103976376 2D 01

ITERATION= 1   TRIAL= 0   F= 0.334877214 4D-03
ALPHA= 0.000000000D 00   DF= -0.525805090 0D-04 |DFV| =   0.725086353 1D-02
CV
   0.208355154 7D 01   0.103454175 2D 01

ITERATION= 1   TRIAL= 1   F= 0.316520733 6D-03
ALPHA= 0.353842335 3D 00   DF= -0.511736799 4D-04 |DFV| =   0.706420490 9D-02
CV
   0.208115942 4D 01   0.103359763 8D 01

ITERATION= 1   TRIAL= 2   F= 0.1607292601D-03
ALPHA= 0.3892265689D 01   DF= -0.3678532412D-04 |DFV|= 0.5278373916D-02
CV
      0.2057238196D 01   0.1024156502D 01

ITERATION= 1   TRIAL= 3   F= 0.4322074203D-04
ALPHA= 0.1647703538D 02   DF= 0.1950894984D-04 |DFV|= 0.6393204957D-02
CV
      0.1972159867D 01   0.9905781234D 00

ITERATION= 2   TRIAL= 0   F= 0.3058857862D-05
ALPHA= 0.0000000000D 00   DF= -0.1760833918D-04 |DFV|= 0.4196229162D-02
CV
      0.2000242079D 01   0.1001661500D 01

ITERATION= 2   TRIAL= 1   F= 0.1108405461D-04
ALPHA= 0.1000000000D 01   DF= 0.3372436775D-04 |DFV|= 0.8039248525D-02
CV
      0.2001781328D 01   0.9977577771D 00

ITERATION= 3   TRIAL= 0   F= 0.2987822515D-07
ALPHA= 0.0000000000D 00   DF= -0.5123322758D-08 |DFV|= 0.7157747106D-04
CV
      0.2000771410D 01   0.1000319052D 01

(continued)

Table 4-8c Continued

```
ITERATION= 3   TRIAL= 1    F=  0.2814233189D-07
ALPHA= 0.3438926638D 00    DF= -0.4972328874D-08   |DFV|=  0.6946839010D-04
CV
     0.2000748664D 01  0.1000309635D 01

ITERATION= 3   TRIAL= 2    F=  0.1363966636D-07
ALPHA= 0.3782781901D 01    DF= -0.3462072219D-08   |DFV|=  0.4837380035D-04
CV
     0.2000521203D 01  0.1000215468D 01

ITERATION= 3   TRIAL= 3    F=  0.2607537759D-08
ALPHA= 0.1510490063D 02    DF=  0.1514342831D-08   |DFV|=  0.2116468886D-04
CV
     0.1999772317D 01  0.9999054368D 00

ITERATION= 4   TRIAL= 0    F=  0.1153806689D-12
ALPHA= 0.0000000000D 00    DF= -0.6724872519D-12   |DFV|=  0.8200532007D-06
CV
     0.2000000106D 01  0.9999997393D 00

ITERATION LIMIT
```

and

$$dy = (Hx)dt + Jd\mu \qquad (4.12)$$

where Brownian motions **u** and μ are zero mean with covariance matrix

$$E\left\{\begin{bmatrix}u(t)\\ \mu(t)\end{bmatrix}\begin{bmatrix}u(\tau)\\ \mu(\tau)\end{bmatrix}'\right\} = \begin{bmatrix}I & K'\\ K & I\end{bmatrix}\min\{t,\tau\}. \qquad (4.13)$$

The class of unbiased filters under consideration is given by

$$dz = [Fz + C(y - Hz)]dt, \; z(0) = E\{x_o\} \qquad (4.14)$$

where **C** is a matrix of coefficients which is optimized so that objective function

$$f = \lim_{t=\infty} E\left\{[x(t) - z(t)]'\,[x(t) - z(t)]\right\} \qquad (4.15)$$

is minimized.

Optimal filter gains can be expressed in terms of the steady-state covariance matrix **X** of error vector $[x(t) - z(t)]$ as

$$C = [G(JK)' + XH']\,(JJ')^{-1}. \qquad (4.16)$$

This covariance matrix satisfies the algebraic Riccati equations given in equation (4.7) when coefficient matrices **P**, **A**, and **R** are properly specified.

FORM2

Computation of coefficient matrices **P**, **A**, and **R** is performed by FORM2 which is summarized as follows:

FORM2 function: Compute

$$A = [F - G(JK)'(JJ')^{-1}H]',$$
$$R = H'(JJ')^{-1}H,$$

and

$$P = G[I - (JK)'(JJ')^{-1}(JK)]G'$$

for use with QUAD.

FORM2 dummy argument list: None.

FORM2 input data list: C1/C2/C3/NU,NMU,NX,NY/FM/GM/HM/JM/KM, where
$1 \leqslant NU \leqslant 3, 1 \leqslant NMU \leqslant 3, 1 \leqslant NX \leqslant 7, 1 \leqslant NY \leqslant 3$.
FM(NX,NX) = **F**.
GM(NX,NU) = **G**.
HM(NY,NX) = **H**.
JM(NY,NMU) = **J**.
KM(NMU,NU) = **K**.

FORM2 output data list: AM/PM/RM, where
AM(NX,NX) = **A**.
PM(NX,NX) = **P**.
RM(NX,NX) = **R**.

Subroutines called by FORM2: MI.

Program FORM2 is listed in Table 4-9. User problems are specified entirely by input data so that a user-supplied subprogram is not needed for the use of FORM2.

Example 5A

An example of the use of program FORM2 for the computation of matrices **P**, **A**, and **R** is the optimal linear filter problem corresponding to

$$\mathbf{F} = \begin{bmatrix} -1 & 0 \\ 0 & -1 \end{bmatrix}, \mathbf{G} = \begin{bmatrix} 1 \\ 0 \end{bmatrix}, \mathbf{H} = [0 \ 1], \mathbf{J} = [1], \text{ and}$$

$$\mathbf{K} = \begin{bmatrix} 1 \\ \frac{1}{2} \end{bmatrix}.$$

A listing of program input cards for this example is given in Table 4-10a, and corresponding program printed output is listed in Table 4-10b.

Example 5B

Computation of covariance matrix **X** for the optimal filter defined in Example 5A can now be accomplished using the Newton method of successive substitutions. Program QUAD can be used with NV = 0 for this purpose. A listing of program input cards for this example is given in Table 4-11a, and corresponding printed output is listed in Table 4-11b.

Table 4-9
Listing of FORM2

```
1       C       FORM2
2       C       REAL*8  AM(NX,NX),FM(NX,NX),GM(NX,NU),HM(NY,NX),JM(NY,NMU),
3       C      1KM(NMU,NU),PM(NX,NX),RM(NX,NX),DDM(NY,NU),DEM(NY,NY),
4       C      2DFM(NX,NY),DGM(NU,NY),DHM(NU,NX),D
5               REAL*8  AM(7,7),FM(7,7),GM(7,3),HM(3,7),JM(3,3),
6              1KM(3,3),PM(7,7),RM(7,7),DDM(3,3),DEM(3,3),
7              2DFM(7,3),DGM(3,3),DHM(3,7)
8               DATA MY/3/
9               REAL*8  D
10              LOGICAL*1 COM(65)
11      1000    FORMAT(65A1)
12      1001    FORMAT(4X,65A1)
13      1002    FORMAT(4D18.10)
14      1003    FORMAT(4I10)
15      1004    FORMAT(4X,4D18.10)
16      1005    FORMAT('1')
17      1006    FORMAT(//,'            NU,NMU,NX,NY')
18      1007    FORMAT('            FM')
19      1008    FORMAT('            GM')
20      1009    FORMAT('            HM')
21      1010    FORMAT('            JM')
22      1011    FORMAT('            KM')
23      1012    FORMAT(//,'          AM')
24      1013    FORMAT('            PM')
25      1014    FORMAT('            RM')
```

(continued)

Table 4-9 Continued

```
26      WRITE (6,1005)
27      DO 1015 I=1,3
28      READ (5,1000) COM
29 1015 WRITE (6,1001) COM
30      READ (5,1003) NU,NMU,NX,NY
31      READ (5,1002) ((FM(I,J),I=1,NX),J=1,NX)
32      READ (5,1002) ((GM(I,J),I=1,NX),J=1,NU)
33      READ (5,1002) ((HM(I,J),I=1,NY),J=1,NX)
34      READ (5,1002) ((JM(I,J),I=1,NY),J=1,NMU)
35      READ (5,1002) ((KM(I,J),I=1,NMU),J=1,NU)
36      WRITE (6,1006)
37      WRITE (6,1003) NU,NMU,NX,NY
38      WRITE (6,1007)
39      WRITE (6,1004) ((FM(I,J),I=1,NX),J=1,NX)
40      WRITE (6,1008)
41      WRITE (6,1004) ((GM(I,J),I=1,NX),J=1,NU)
42      WRITE (6,1009)
43      WRITE (6,1004) ((HM(I,J),I=1,NY),J=1,NX)
44      WRITE (6,1010)
45      WRITE (6,1004) ((JM(I,J),I=1,NY),J=1,NMU)
46      WRITE (6,1011)
47      WRITE (6,1004) ((KM(I,J),I=1,NMU),J=1,NU)
48      DO 1019 I=1,NY
49      DO 1017 J=I,NY
50      D=0.D0
```

(continued)

```
51         DO 1016 K=1,NMU
52  1016   D=D+JM(I,K)*JM(J,K)
53         DEM(I,J)=D
54  1017   DEM(J,I)=D
55         DO 1019 J=1,NU
56         D=0.D0
57         DO 1018 K=1,NMU
58  1018   D=D+JM(I,K)*KM(K,J)
59  1019   DDM(I,J)=D
60         CALL MI(MY,NY,DEM)
61         DO 1023 I=1,NU
62         DO 1021 J=1,NY
63         D=0.D0
64         DO 1020 K=1,NY
65  1020   D=D+DDM(K,I)*DEM(K,J)
66  1021   DGM(I,J)=D
67         DO 1023 J=1,NX
68         D=0.D0
69         DO 1022 K=1,NY
70  1022   D=D+DEM(I,K)*HM(K,J)
71  1023   DHM(I,J)=D
72         DO 1027 I=1,NX
73         DO 1025 J=1,NY
74         D=0.D0
75         DO 1024 K=1,NU
76  1024   D=D+GM(I,K)*DGM(K,J)
77  1025   DFM(I,J)=D
```

Table 4-9 Continued

```
78            DO 1027 J=I,NX
79            D=0.D0
80            DO 1026 K=1,NY
81   1026     D=D+HM(K,I)*DHM(K,J)
82   1027     RM(I,J)=D
83            DO 1029 I=1,NX
84            DO 1029 J=1,NU
85            D=GM(I,J)
86            DO 1028 K=1,NY
87   1028     D=D-DFM(I,K)*DDM(K,J)
88   1029     DHM(J,I)=D
89            DO 1033 I=1,NX
90            DO 1031 J=1,NX
91            D=FM(I,J)
92            DO 1030 K=1,NY
93   1030     D=D-DFM(I,K)*HM(K,J)
94   1031     AM(J,I)=D
95            DO 1033 J=I,NX
96            D=0.D0
97            DO 1032 K=1,NU
98   1032     D=D+DHM(K,I)*GM(J,K)
99   1033     PM(I,J)=D
100           WRITE (6,1012)
101           WRITE (6,1004) ((AM(I,J),I=1,NX),J=1,NX)
102           WRITE (6,1013)
```

```
103       WRITE (6,1004)  ((PM(I,J),I=1,J),J=1,NX)
104       WRITE (6,1014)
105       WRITE (6,1004)  ((RM(I,J),I=1,J),J=1,NX)
106       WRITE (6,1005)
107       WRITE (7,1002)  ((AM(I,J),I=1,NX),J=1,NX)
108       WRITE (7,1002)  ((PM(I,J),I=1,J),J=1,NX)
109       WRITE (7,1002)  ((RM(I,J),I=1,J),J=1,NX)
110       STOP
111       END
```

Table 4-10a
Listing of User-supplied Input Data for Example 5A

```
1  FORM2(MI)   EXAMPLE 5A
2  MERRIAM
3  UNIVERSITY OF ROCHESTER
4              1                  2
5  -1.            1.      0.   0.       1.     -1.
6   1.                    1.
7   0.
8   1.
9   .5
```

Table 4-10b
Printed Output from FORM2 for Example 5A

```
FORM2(MI)   EXAMPLE 5A
MERRIAM
UNIVERSITY OF ROCHESTER


NU,NMU,NX,NY
    1    1    2    1
FM
    -0.1000000000D 01   0.0000000000D 00   0.0000000000D 00  -0.1000000000D 01
GM
     0.1000000000D 01   0.0000000000D 00
HM
     0.0000000000D 00   0.1000000000D 01
JM
     0.1000000000D 01
KM
     0.5000000000D 00
AM
    -0.1000000000D 01  -0.5000000000D 00  -0.0000000000D 00  -0.1000000000D 01
PM
     0.7500000000D 00   0.0000000000D 00   0.0000000000D 00
RM
     0.0000000000D 00   0.0000000000D 00   0.1000000000D 01
```

Table 4–11a
Listing of User-supplied Input Data for Example 5B

```
1  QUAD(IXM,INC,LIN2,MI)     EXAMPLE 5B
2  MERRIAM
3  UNIVERSITY OF ROCHESTER
4         2            2            2            0
5  0.
6  -1.                    -.5                    0.
7  .75            0.                   0.       -1.
8  0.            0.                              0.
9  0.
```

Table 4-11b
Printed Output from QUAD for Example 5B

```
QUAD(DXM,INC,LIN2,MI)   EXAMPLE 5B
MERRIAM
UNIVERSITY OF ROCHESTER

NI,MODE,NX,NV
 2       2       2       0
DXMIN
 0.000000000D 00
AM
-0.100000000D 01  -0.500000000D 00   0.000000000D 00  -0.100000000D 01
PM
 0.750000000D 00   0.000000000D 00   0.000000000D 00
RM
 0.000000000D 00   0.000000000D 00   0.100000000D 01
XM
 0.000000000D 00   0.000000000D 00   0.000000000D 00

ITERATION= 1   ||DXM||=  0.750000000D 00
XM
 0.000000000D 00   0.000000000D 00   0.000000000D 00

ITERATION= 2   ||DXM||=  0.138777878 1D-16
XM
 0.375000000D 00   0.000000000D 00   0.000000000D 00

TRIAL LIMIT
```

GAIN2

Computation of the gain matrix for linear optimal filters can finally be accomplished using program GAIN2 which is summarized as follows:

GAIN2 function: Compute

$$C = [G(JK)' + XH'](JJ')^{-1}.$$

GAIN2 dummy argument list: None.

GAIN2 input data list: C1/C2/C3/NU,NMU,NX,NY/GM/HM/JM/KM/XM, where
    $1 \leqslant NU \leqslant 3, 1 \leqslant NMU \leqslant 3, 1 \leqslant NX \leqslant 7, 1 \leqslant NY \leqslant 3$.
    GM(NX,NU) = G.
    HM(NY,NX) = H.
    JM(NY,NMU) = J.
    KM(NMU,NU) = K.
    XM(NX,NX) = X.

GAIN2 output data list: CM, where
    CM(NX,NY) = C.

Subprograms called by GAIN2: MI.

    Program GAIN2 is listed in Table 4-12. No user-supplied subprograms are required for the use of GAIN2.

Example 5C

Program GAIN2 is illustrated using the linear optimal filter problem specified in Example 5A. Program input cards are listed in Table 4-13a, and corresponding program printed output from GAIN2 is listed in Table 4-13b.

4.3 Simulation of Human Controller Models Based on Optimization

Stochastic models have been proposed to represent human controllers which operate in a compensatory tracking mode. These models include adaptation in the form of parameters which are adjusted to minimize suitably defined tracking error. These models also include multiplicative stochastic disturbances which simulate human observation noise and human motor noise.

Table 4-12
Listing of GAIN2

```
1      C      GAIN2
2      C      REAL*8 CM(NX,NY),GM(NX,NU),HM(NY,NX),JM(NY,NMU),KM(NMU,NU),
3      C     1XM(NX,NX),DDM(NX,NY),DEM(NY,NY),DFM(NY,NU),D
4             REAL*8 CM(7,3),GM(7,3),HM(3,7),JM(3,3),KM(3,3),
5            1XM(7,7),DDM(7,3),DEM(3,3),DFM(3,3)
6             DATA MY/3/
7             REAL*8 D
8             LOGICAL*1 COM(65)
9      1000   FORMAT(65A1)
10     1001   FORMAT(4X,65A1)
11     1002   FORMAT(4D18.10)
12     1003   FORMAT (4I10)
13     1004   FORMAT(4X,4D18.10)
14     1005   FORMAT('1')
15     1006   FORMAT(//,'       NU,NMU,NX,NY')
16     1007   FORMAT('            GM')
17     1008   FORMAT('            HM')
18     1009   FORMAT('            JM')
19     1010   FORMAT('            KM')
20     1011   FORMAT('            XM')
21     1012   FORMAT(//,'         CM')
22             WRITE (6,1005)
23             DO 1013 I=1,3
24             READ (5,1000) COM
```

Table 4-12 Continued

```
25  1013  WRITE (6,1001) COM
26        READ  (5,1003) NU,NMU,NX,NY
27        READ  (5,1002) ((GM(I,J),I=1,NX),J=1,NU)
28        READ  (5,1002) ((HM(I,J),I=1,NY),J=1,NX)
29        READ  (5,1002) ((JM(I,J),I=1,NY),J=1,NMU)
30        READ  (5,1002) ((KM(I,J),I=1,NMU),J=1,NU)
31        READ  (5,1002) ((XM(I,J),I=1,J),J=1,NX)
32        WRITE (6,1006)
33        WRITE (6,1003) NU,NMU,NX,NY
34        WRITE (6,1007)
35        WRITE (6,1004) ((GM(I,J),I=1,NX),J=1,NU)
36        WRITE (6,1008)
37        WRITE (6,1004) ((HM(I,J),I=1,NY),J=1,NX)
38        WRITE (6,1009)
39        WRITE (6,1004) ((JM(I,J),I=1,NY),J=1,NMU)
40        WRITE (6,1010)
41        WRITE (6,1004) ((KM(I,J),I=1,NMU),J=1,NU)
42        WRITE (6,1011)
43        WRITE (6,1004) ((XM(I,J),I=1,J),J=1,NX)
44        DO 1014 I=1,NX
45        DO 1014 J=I,NX
46  1014  XM(J,I)=XM(I,J)
47        DO 1016 I=1,NY
48        DO 1016 J=I,NY
49        D=0.D0
50        DO 1015 K=1,NMU
```

```
51    1015  D=D+JM(I,K)*JM(J,K)
52          DEM(I,J)=D

53    1016  DEM(J,I)=D
54          CALL MI(MY,NY,DEM)
55          DO 1018 I=1,NY
56          DO 1018 J=1,NU
57          D=0.D0
58          DO 1017 K=1,NMU
59    1017  D=D+JM(I,K)*KM(K,J)
60    1018  DFM(I,J)=D
61          DO 1021 I=1,NX
62          DO 1021 J=1,NY
63          D=0.D0
64          DO 1019 K=1,NU
65    1019  D=D+GM(I,K)*DFM(J,K)
66          DO 1020 K=1,NX
67    1020  D=D+XM(I,K)*HM(J,K)
68    1021  DDM(I,J)=D
69          DO 1023 I=1,NX
70          DO 1023 J=1,NY
71          D=0.D0
72          DO 1022 K=1,NY
73    1022  D=D+DDM(I,K)*DEM(K,J)
74    1023  CM(I,J)=D
75          WRITE (6,1012)
76          WRITE (6,1004) ((CM(I,J),I=1,NX),J=1,NY)
77          WRITE (6,1005)
78          WRITE (7,1002) ((CM(I,J),I=1,NX),J=1,NY)
79          STOP
80          END
```

Table 4-13a
Listing of User-supplied Input for Example 5C

```
1   GAIN2(MI)      EXAMPLE 5C
2   MERRIAM
3   UNIVERSITY OF ROCHESTER
4              1              1             2            1
5       1.                    0.
6       0.                    1.
7       1.
8       .5
9       .375                  0.                         0.
```

Description of the Basic Method

Stochastic equations for human controller simulation are taken to be

$$d\mathbf{x} = (\mathbf{A}\mathbf{x})dt + d\mathbf{u} + \sum_{k=1}^{m} (\mathbf{B}_k \mathbf{x})dw_k, \mathbf{x}(0) = \mathbf{x}_o \qquad (4.17)$$

where vectors $\mathbf{u}(t)$ and $\mathbf{w}(t)$ denote zero-mean, independent Brownian motions which have covariance matrices

$$E\{\mathbf{u}(t)\mathbf{u}'(\tau)\} = \mathbf{C} \min\{t, \tau\} \text{ and } E\{\mathbf{w}(t)\mathbf{w}'(\tau)\} = \mathbf{I} \min\{t, \tau\}. \qquad (4.18)$$

The corresponding steady-state covariance matrix $\mathbf{X}$ of state vector $\mathbf{x}(t)$ satisfies

$$0 = \mathbf{X}\mathbf{A}' + \mathbf{A}\mathbf{X} + \sum_{k=1}^{m} \mathbf{B}_k \mathbf{X} \mathbf{B}_k' + \mathbf{C}. \qquad (4.19)$$

The existence of positive semidefinite solutions to these modified algebraic Lyapunov equations depends upon the properties of coefficient matrices $\mathbf{A}, \mathbf{B}_k$, and $\mathbf{C}$ in a complicated way which is described in chapter 5 of *Automated Design of Control Systems* [1].

Solution of equation (4.19) by Newton's method of successive substitutions requires the selection of a positive semidefinite initialization $\mathbf{X}_o$ which also meets other complicated criteria. Meeting these criteria a priori may be next to impossible. However, these criteria can be met in a routine fashion if the algebraic equations given in equation (4.19) are first converted into the

Table 4-13b
Printed Output from GAIN2 for Example 5C

```
GAIN2(MI)   EXAMPLE  5C
MERRIAM
UNIVERSITY OF ROCHESTER

NU, NMU, NX, NY
     1         1         2         1
GM
   0.1000000000D 01   0.0000000000D 00
HM
   0.0000000000D 00   0.1000000000D 01
JM
   0.1000000000D 01
KM
   0.5000000000D 00
XM
   0.3750000000D 00   0.0000000000D 00   0.0000000000D 00
CM
   0.5000000000D 00   0.0000000000D 00
```

matrix differential equation that corresponds to the transient covariance matrix of state vector $x(t)$ and then standard numerical integration techniques are employed. The first set of programs and subprograms described in this section are included in the set of system-supplied programs primarily for this purpose. For example, the Gill modification [8] of the Runge-Kutta method of numerical integration is used in the subsequent program presented in this section.

Minimization of objective function $F(c) = f[X(c)]$ with respect to a parameter vector c requires computation of gradient vector $f = F_c$ where each vector element is of the form

$$f_i = tr\{f_X X_{c_i}\}. \tag{4.20}$$

Covariance matrix X derives its dependence on vector c via equation (4.19) and the dependence of coefficient matrices A, B_k, C on vector c. Methods of successive approximation, which are implemented in either subprograms MIN1 or MIN3, can be applied directly to this minimization problem.

INT, INTS, and DXM

Numerical integration of a set of first-order, ordinary differential equations is implemented by subprogram INT which is summarized as follows:

INT function: Integrate

$$\dot{x} = f(x,t); x(t_o) = x_o.$$

INT dummy argument list: None.

INT input data list: C1/C2/C3/NII, NP,NX/DT/T/XV, where
  NII = number of integration increments.
  NP = number of integration increments omitted between output.
  $1 \leqslant NX \leqslant 28$.
  DT = length of integration increments.
  T = t_o.
  XV(NX) = x_o.

INT output data list: XV, where
  XV(NX) = x_f.
Subprograms called by INT: INTS for the computation of

  DXV(NX) = $f(x,t)$.

OPTIMIZATION ON AN INFINITE TIME INTERVAL

Subprogram INT is listed in Table 4-14. User problems are specified for subprogram INT by writing subprogram INTS. This subprogram is specified as follows:

INTS function: Compute $f(x,t)$ for use with INT.

INTS dummy argument list: MX,T,XV,DXV,MODE where
    $T = t$.
    XV(MX) = x.
    DXV(MX) = $f(x,t)$.
    MODE = 1 for program initialization and 2 otherwise.

Subprogram INTS can be further specialized easily by employing a subprogram to form appropriate derivatives for the particular application of numerical integration discussed previously. This additional subprogram, which is also used in other examples, is summarized as follows:

DXM function: Compute

$$\dot{X} = XA + A'X + \sum_{k=1}^{m} B'_k XB_k + P - XRX$$

as a subprogram.

DXM dummy argument list: MX,MV,NX,NV,AM,BA,PM,RM,XM,DXM, where
    $1 \leqslant MX \leqslant 7, 1 \leqslant MV \leqslant 3, 1 \leqslant NX \leqslant MX, 0 \leqslant NV \leqslant MV$.
    AM(MX,MX) = **A**.
    BA(MV,MX,MX) = $B_1/B_2/\ldots$.
    PM(MX,MX) = **P**
    RM(MX,MX) = **R**.
    XM(MX,MX) = **X**.
    DXM(MX,MX) = $\dot{X}$.

DXM input data list: None.

DXM output data list: None.

Subprograms called by DXM: None

Subprogram DXM is listed in Table 4-15.

Table 4-14
Listing of INT

```
1            SUBROUTINE INT
2     C      REAL*8 XV(NX),DXV(NX),DV(NX),T,DT,DTH,G1,G2,G3,G4,G5,G6,G7,
3     C     1G8,D,DSQRT,DABS
4            REAL*8 XV(28),DXV(28),DV(28)
5            DATA MX/28/
6            DATA DXV/28*0.D0/
7            REAL*8 T,DT,DTH,G1,G2,G3,G4,G5,G6,G7,G8,D,DSQRT,DABS
8            LOGICAL*1 COM(65)
9     1000   FORMAT(65A1)
10    1001   FORMAT(4X,65A1)
11    1002   FORMAT(4D18.10)
12    1003   FORMAT(3I10)
13    1004   FORMAT(4X,4D18.10)
14    1005   FORMAT(' 1')
15    1006   FORMAT(//,'           NII,NP,NX')
16    1007   FORMAT('           DT')
17    1008   FORMAT(//,'           STEP=',I4,'      T=',D18.10,'      ||DXV||=',D18.10)
18    1009   FORMAT('           XV')
19           WRITE (6,1005)
20           DO 1010 I=1,3
21           READ (5,1000) COM
22    1010   WRITE (6,1001) COM
23           READ (5,1003) NI,NP,NX
24           READ (5,1002) DT
```

(continued)

```
25        READ (5,1002)  T
26        READ (5,1002) (XV(I),I=1,NX)
27        WRITE (6,1006)
28        WRITE (6,1003)  NI,NP,NX
29        WRITE (6,1007)
30        WRITE (6,1004)  DT
31        D=DSQRT(5.D-1)
32        G1=DT*(1.D0-D)
33        G2=DT*(1.D0+D)
34        D=3.D0*D
35        G6=D-2.D0
36        G8=-(D+2.D0)
37        G3=DT/6.D0
38        G4=-DT/3.D0
39        D=DSQRT(2.D0)
40        G5=2.D0-D
41        G7=2.D0+D
42        L=0
43        NIP=NP
44        DTH=5.D-1*DT
45        NIC=0
46        CALL INTS(MX,T,XV,DXV,1)
47 1011   CALL INTS(MX,T,XV,DXV,2)
48        NIP=NIP+1
49        IF (NIP .LE. NP) GO TO 1013
50        D=0.D0
51        DO 1012 I=1,NX
52        IF (D .GE. DABS(DXV(I))) GO TO 1012
```

Table 4-14 Continued

```
53          D=DABS(DXV(I))
54   1012 CONTINUE
55        WRITE (6,1008) NIC,T,D
56        WRITE (6,1009)
57        WRITE (6,1004) (XV(I),I=1,NX)
58        NIP=0
59        IF (L .EQ. 1) GO TO 1018
60   1013 DO 1014 I=1,NX
61        DV(I)=DXV(I)
62   1014 XV(I)=XV(I)+DTH*DXV(I)
63        T=T+DTH
64        CALL INTS(MX,T,XV,DXV,2)
65        DO 1015 I=1,NX
66        XV(I)=XV(I)+G1*(DXV(I)-DV(I))
67   1015 DV(I)=G5*DXV(I)+G6*DV(I)
68        CALL INTS(MX,T,XV,DXV,2)
69        DO 1016 I=1,NX
70        XV(I)=XV(I)+G2*(DXV(I)-DV(I))
71   1016 DV(I)=G7*DXV(I)+G8*DV(I)
72        T=T+DTH
```

```
73          CALL INTS(MX,T,XV,DXV,2)
74          DO 1017 I=1,NX
75   1017   XV(I)=XV(I)+G3*DXV(I)+G4*DV(I)
76          NIC=NIC+1
77          IF (NIC .LT. NI) GO TO 1011
78          L=1
79          NIP=NP
80          GO TO 1011
81   1018   WRITE (6,1005)
82          WRITE (7,1002) (XV(I),I=1,NX)
83          RETURN
84          END
```

Table 4-15
Listing of DXM

```
        SUBROUTINE DXM(MX,MV,NX,NV,AM,BA,PM,RM,XM,DYM)
C       COMPUTATION OF DXM
        REAL*8 AM(NX,NX),BA(NV,NX,NX),PM(NX,NX),RM(NX,NX),
       1XM(NX,NX),DYM(NX,NX),YA(NV,NX,NX),YM(NX,NX),D
        REAL*8 AM(MX,MX),BA(MV,MX,MX),PM(MX,MX),RM(MX,MX),
       1XM(MX,MX),DYM(MX,MX),D
        REAL*8 YA(3,7,7),YM(7,7)
        DO 1005 I=1,NX
        DO 1005 J=1,NX
        IF (NV) 1003,1003,1000
 1000   DO 1002 K=1,NV
        D=0.D0
        DO 1001 L=1,NX
 1001   D=D+XM(I,L)*BA(K,L,J)
 1002   YA(K,I,J)=D
 1003   D=0.D0
        DO 1004 K=1,NX
 1004   D=D+RM(I,K)*XM(K,J)
 1005   YM(I,J)=D
        DO 1010 J=1,NX
        DO 1010 I=1,J
```

```
22            D=PM(I,J)
23            IF (NV) 1008,1008,1006
24   1006     DO 1007 K=1,NV
25            DO 1007 L=1,NX
26   1007     D=D+BA(K,L,I)*YA(K,L,J)
27   1008     DO 1009 K=1,NX
28   1009     D=D+XM(I,K)*(AM(K,J)-YM(K,J))+AM(K,I)*XM(K,J)
29            DYM(I,J)=D
30   1010     DYM(J,I)=D
31            RETURN
32            END
```

Example 6A

An elementary example of simulating the human tracking problem is to find the optimal parameter c_1 for matrices

$$A = \begin{bmatrix} 0 & -c_1 \\ 1 & -1 \end{bmatrix}, B_1 = \begin{bmatrix} 0 & 0 \\ 1 & 0 \end{bmatrix}, \text{ and } C = \begin{bmatrix} 1 & 0 \\ 0 & 0 \end{bmatrix}$$

which appear in equations (4.17) and (4.19). An initial trial for covariance matrix X can be obtained by numerical integration by first writing subprogram INTS to call subprogram DXM in an appropriate fashion. Subprogram INTS is supplied by the user for this purpose and is listed in Table 4-16a. A listing of program input cards for this example is given in Table 4-16b, and corresponding program output is listed in Table 4-16c. Numerical integration yields matrix

$$\begin{bmatrix} 1.90 & 0.94 \\ 0.94 & 1.85 \end{bmatrix}$$

at approximately 5.0 seconds as an approximation to covariance matrix X for parameter $c_1 = 0.5$.

LOAD2

Subprogram LOAD2 is used to specialize either MIN1 or MIN3 to minimizing the objective function $f[X(c)]$ with respect to vector c. This subprogram is summarized as follows:

LOAD2 function: Compute the solution to

$$XA' + AX + \sum_{k=1}^{m} B_k XB'_k + C = 0 \;,$$

$$X_{c_i}A' + AX_{c_i} + \sum_{k=1}^{m} B_k X_{c_i} B'_k + \{C_{c_i} + XA'_{c_i} + A_{c_i}X$$

$$+ \sum_{k=1}^{m} [(B_k)_{c_i} XB'_k + B_k X(B'_k)_{c_i}]\} = 0 \;,$$

and

$$f_i = tr\{f_X X_{c_i}\}$$

as a subprogram for use with MIN1S or MIN3S.

LOAD2 dummy argument list: MC,NC,NV,NX,AM,BA,CM,DAA,DBAA,DCA, DFM,XM,DXA,DFV, where
   $1 \leqslant MC \leqslant 21, 1 \leqslant NC \leqslant MC, 0 \leqslant NV \leqslant 3, 1 \leqslant NX \leqslant 7$.
   AM(NX,NX) = **A**.
   BA(NV,NX,NX) = $\mathbf{B}_1/\mathbf{B}_2/\ldots$.
   CM(NX,NX) = **C**.
   DAA(NC,NX,NX) = $\mathbf{A}_{c_1}/\mathbf{A}_{c_2}/\ldots$.
   DBAA(NC,NV,NX,NX) = $(\mathbf{B}_1/\mathbf{B}_2/\ldots)_{c_1}/(\mathbf{B}_1/\mathbf{B}_2/\ldots)_{c_2}/\ldots$.
   DCA(NC,NX,NX) = $\mathbf{C}_{c_1}/\mathbf{C}_{c_2}/\ldots$.
   DFM(NX,NX) = f_X.
   XM(NX,NX) = **X**.
   DXA(NC,NX,NX) = $\mathbf{X}_{c_1}/\mathbf{X}_{c_2}/\ldots$.
   DFV(MC) = f.

LOAD2 input data list: None.

LOAD2 output data list: None.

Subprograms called by LOAD2: INC,LIN2,MI.

Subprogram LOAD2 is listed in Table 4-17. This subprogram calls subprogram LIN2 for the solution of matrix algebraic equations listed above by inverting their equivalent coefficient matrix. Subprogram LIN2 is discussed in chapter 6. Either MIN1C and MIN1S or MIN3C and MIN3S must be supplied by the user in order to employ subprogram LOAD2 with either subprogram MIN1 or MIN3, respectively.

Example 6B

Subprogram LOAD2 is illustrated using the simulation problem defined in Example 6A and objective function $f = x_{11}$. Program MIN1C and subprogram MIN1S are supplied by the user and are listed in Table 4-18a. Program input cards, which are also supplied by the user, are listed in Table 4-18b, and corresponding program printed outputs from LOAD2 and MIN1 are listed in Table 4-18c.

Table 4-16a
Listing of User-supplied Programs for Example 6A

```
1        C    INTC    EXAMPLE 6A
2                     CALL INT
3                     STOP
4                     END
5                     SUBROUTINE INTS(MX,T,XV,DXV,MODE)
6                     REAL*8 XV(MX),DXV(MX),T
7                     REAL*8 AM(2,2),BA(1,2,2),CM(2,2),DM(2,2),XM(2,2),YXM(2,2),C1
8                     DATA AM/4*0.D0/,BA/4*0.D0/,CM/4*0.D0/,DM/4*0.D0/
9        1000         FORMAT(D18.10)
10       1001         FORMAT(//,4X,'CV(1)=',D18.10)
11                    GO TO (1002,1003),MODE
12       1002         READ (5,1000) C1
13                    WRITE (6,1001) C1
14                    AM(1,2)=1.D0
15                    AM(2,1)=-C1
16                    AM(2,2)=-1.D0
17                    BA(1,1,2)=1.D0
18                    CM(1,1)=1.D0
19                    RETURN
20       1003         XM(1,1)=XV(1)
```

```
21      XM(2,1) = XV(2)
22      XM(1,2) = XV(2)
23      XM(2,2) = XV(3)
24      CALL DXM(2,1,2,1,AM,BA,CM,DM,XM,YXM)
25      DXV(1) = YXM(1,1)
26      DXV(2) = YXM(1,2)
27      DXV(3) = YXM(2,2)
28      RETURN
29      END
```

Table 4-16b
Listing of User-supplied Input Data for Example 6A

```
1    INT(DXM,INTS)    EXAMPLE 6A
2    MERRIAM
3    UNIVERSITY OF ROCHESTER
4             500            49           3
5    .01
6    0.
7    0.                  0.                              0.
8    .5
```

Table 4-16c
Printed Output from INT and INTS for Example 6A

```
INT(DXM,INTS)    EXAMPLE 6A
MERRIAM
UNIVERSITY OF ROCHESTER

NII,NP,NX
 500        49         3
DT
  0.1000000000D-01

CV(1) =  0.5000000000D 00

STEP=   0    T=  0.0000000000D 00    ||DXV|| =  0.1000000000D 01
XV
  0.0000000000D 00  0.0000000000D 00  0.0000000000D 00

STEP=  50    T=  0.5000000000D 00    ||DXV|| =  0.9043366419D 00
XV
  0.4828899158D 00  0.9566335811D-01  0.1169987319D 00

STEP= 100    T=  0.1000000000D 01    ||DXV|| =  0.7139066822D 00
XV
  0.8889307213D 00  0.2860933178D 00  0.4016097559D 00
```

(continued)

Table 4-16c Continued

```
STEP= 150    T=  0.1500000000D 01    ||DXV||=  0.6703879096D 00
XV
    0.1197028938D 01   0.4770380748D 00   0.7403585892D 00

STEP= 200    T=  0.2000000000D 01    ||DXV||=  0.5715353210D 00
XV
    0.1418377701D 01   0.6301450468D 00   0.1053566237D 01

STEP= 250    T=  0.2500000000D 01    ||DXV||=  0.4405593144D 00
XV
    0.1574171053D 01   0.7399481596D 00   0.1306754029D 01

STEP= 300    T=  0.3000000000D 01    ||DXV||=  0.3211222464D 00
XV
    0.1684294168D 01   0.8146225618D 00   0.1496208523D 01

STEP= 350    T=  0.3500000000D 01    ||DXV||=  0.2285360173D 00
XV
    0.1763600716D 01   0.8649026252D 00   0.1632434974D 01
```

STEP= 400   T= 0.4000000000D 01 ||DXV||= 0.1625853299D 00
XV 0.1821986220D 01   0.8995121086D 00   0.1729212554D 01

STEP= 450   T= 0.4500000000D 01 ||DXV||= 0.1173014327D 00
XV 0.1865719707D 01   0.9242561283D 00   0.1798465265D 01

STEP= 500   T= 0.5000000000D 01 ||DXV||= 0.8623148203D-01
XV 0.1898789590D 01   0.9425945827D 00   0.1848873637D 01

Table 4-17
Listing of LOAD 2

```
1            SUBROUTINE LOAD2(MC,NC,NV,NX,AM,BA,CM,DAA,DBAA,DCA,DFM,
2           1XM,DXA,DFV)
3     C     SUBPROGRAM FOR MIN1 AND MIN3
4           REAL*8 AM(NX,NX),BA(NV,NX),CM(NX,NX),DAA(NC,NX,NX),
5          1DBAA(NC,NV,NX,NX),DCA(NC,NX,NX),DFM(NX,NX),DFV(NC),
6          2BM(NX,NX),XM(NX,NX),YM(NT,NT),YA(NV,NX,NX),DXM(NX,NX),
7          3DXA(NC,NX,NX),D
8           REAL*8 AM(NX,NX),BA(NV,NX,NX),CM(NX,NX),DAA(NC,NX,NX),
9          1DBAA(NC,NV,NX,NX),DCA(NC,NX,NX),DFM(NX,NX),DFV(MC),
10         2XM(NX,NX),DXA(NC,NX,NX),D
11          REAL*8 BM(7,7),YM(28,28),YA(3,7,7),DXM(7,7)
12          DATA MT/28/,MX/7/,MV/3/
13          DO 1003 I=1,NX
14          DO 1002 J=1,NX
15          IF (NV) 1002,1002,1000
16     1000 DO 1001 K=1,NV
17     1001 YA(K,J,I)=BA(K,I,J)
18     1002 BM(J,I)=AM(I,J)
19          DO 1003 J=I,NX
20     1003 DFM(J,I)=DFM(I,J)
21          CALL INC(MT,MX,MV,NX,NV,BM,YA,YM)
22          CALL LIN2(MT,NX,NX,YM,CM,XM)
23          IF (NV) 1007,1007,1004
24     1004 DO 1006 L=1,NV
```

(continued)

```
25         DO 1006 I=1,NX
26         DO 1006 J=1,NX
27         D=0.D0
28         DO 1005 K=1,NX
29  1005   D=D+XM(I,K)*BA(L,J,K)
30  1006   YA(L,I,J)=D
31  1007   DO 1018 L=1,NC
32         DO 1008 J=1,NX
33         DO 1008 I=1,J
34  1008   DXM(I,J)=DCA(L,I,J)
35         IF (NV) 1013,1013,1009
36  1009   DO 1012 K=1,NV
37         DO 1011 I=1,NX
38         DO 1011 J=1,NX
39         D=0.D0
40         DO 1010 KI=1,NX
41  1010   D=D+DBAA(L,K,I,KI)*YA(K,KI,J)
42  1011   BM(I,J)=D
43         DO 1012 I=1,NX
44         DO 1012 J=I,NX
45  1012   DXM(I,J)=DXM(I,J)+BM(I,J)+BM(J,I)
46  1013   DO 1015 I=1,NX
47         DO 1015 J=1,NX
48         D=0.D0
49         DO 1014 K=1,NX
50  1014   D=D+DAA(L,I,K)*XM(K,J)
51  1015   BM(I,J)=D
52         DO 1016 I=1,NX
```

Table 4-17 Continued

```
53            DO   1016   J=I,NX
54     1016   DXM(I,J)=DXM(I,J)+BM(I,J)+BM(J,I)
55            CALL LIN2(MT,MX,NX,YM,DXM,DXM)
56            D=0.D0
57            DO   1017   I=1,NX
58            DO   1017   K=1,NX
59            DXA(L,I,K)=DXM(I,K)
60     1017   D=D+DFM(I,K)*DXM(K,I)
61     1018   DFV(L)=D
62            RETURN
63            END
```

Table 4-18a
Listing of User-supplied Programs for Example 6B

```
1     C     MIN1C   EXAMPLE 6B
2           CALL MIN1
3           STOP
4           END
5           SUBROUTINE MIN1S(MC,CV,F,DFV,MODE)
6           REAL*8 CV(MC),DFV(MC),F
7           REAL*8 AM(2,2),BA(1,2,2),CM(2,2),DAA(1,2,2),
8          1DBAA(1,1,2,2),DCA(1,2,2),DFM(2,2),XM(2,2),DXA(1,2,2)
9           DATA AM/0D0,1D0,0D0,-1D0/
10          DATA BA/0D0,1D0,0D0,0D0/
11          DATA CM/1D0,3*0D0/
12          DATA DAA/0D0,0D0,-1D0,0D0/
13          DATA DBAA/4*0D0/
14          DATA DCA/4*0D0/
15          DATA DFM/1D0,3*0D0/
16          GO TO (1001,1000),MODE
17    1000  AM(1,2)=-CV(1)
18          CALL LOAD2(21,1,1,2,AM,BA,CM,DAA,DBAA,DCA,DFM,XM,DXA,DFV)
19          F=XM(1,1)
20    1001  RETURN
21          END
```

Table 4-18b
Listing of User-supplied Input Data for Example 6B

```
1  MIN1(INC,LIN2,LOAD2,MI,MIN1S)   EXAMPLE 6B
2  MERRIAM
3  UNIVERSITY OF ROCHESTER
4            1            6         0         1
5  .01                    0.                  0.       10.
6  .1                     1.                  .5       2.
7  .5
```

Table 4-18c
Printed Output from LOAD2 and MIN1 for Example 6B

```
MIN1(INC,LIN2,LOAD2,MI,MIN1S)     EXAMPLE 6B
MERRIAM
UNIVERSITY OF ROCHESTER

NI,NT,NR,NC
 1      6          0         1
RATIO,DFMIN,FMIN,FMAX
 0.1000000000D-01  0.0000000000D 00  0.0000000000D 00  0.1000000000D 02
ALMIN,ALMAX,RMIN,RMAX
 0.1000000000D 00  0.1000000000D 01  0.5000000000D 01  0.2000000000D 01

ITERATION=  0    TRIAL=  0       F=  0.2000000000D 01
ALPHA=  0.0000000000D 00        DF= -0.1777777778D 01    |DFV|=  0.1333333333D 01
CV
 0.5000000000D 00

ITERATION=  0    TRIAL=  1       F=  0.9272727273D 01
ALPHA=  0.1000000000D 01        DF=  0.7180165289D 02    |DFV|=  0.5385123967D 02
CV
 0.1833333333D 01
```

(continued)

Table 4-18c Continued

```
ITERATION= 0   TRIAL= 2    F=  0.244496246 9D 01
ALPHA=  0.5764397244D 00   DF=  0.3324299430D 01   |DFV|=  0.2493224573D 01
CV
     0.1268586299D 01

ITERATION= 0   TRIAL= 3    F=  0.1868585956D 01
ALPHA=  0.2010630105D 00   DF=  0.1878221403D 00   |DFV|=  0.1408666052D 00
CV
     0.7680840140D 00

ITERATION= 0   TRIAL= 4    F=  0.1866064458D 01
ALPHA=  0.1707367030D 00   DF= -0.2368954229D-01   |DFV|=  0.1776715672D-01
CV
     0.7276489373D 00

ITERATION= 1   TRIAL= 0    F=  0.1866025404D 01
ALPHA=  0.0000000000D 00   DF= -0.7145426873D-09   |DFV|=  0.2673093128D-04
CV
     0.7320574559D 00

ITERATION LIMIT
```

5
Optimization with Functionals Defined on a Finite Time Interval

Computer programs presented in this chapter are intended to be representative of computational techniques that are implemented for solving optimization problems which are defined on a finite time interval. Examples of both parameter optimization and forcing-function optimization are included. The derivations which pertain to these examples can be found in chapter 6 of *Automated Design of Control Systems* [1].

5.1 Optimal Control Parameters with a Fixed Time Interval

Parameter optimization arises frequently in optimal control and filter problems defined on a finite time interval even though forcing-function optimization is often associated with such problems. The computer programs described in this section apply to a class of control problems, and these programs illustrate an advanced application of the methods of successive approximations discussed in chapter 3.

Description of the Basic Method

The class of control optimization problems is defined as the minimization of performance functional

$$J = \int_0^{t_f} F(\mathbf{x}, \mathbf{c}, t)dt + G(\mathbf{x}_f, \mathbf{c}) \quad (5.1)$$

with respect to parameter vector $\mathbf{c}$ where final time t_f is fixed. State vector $\mathbf{x}(t)$ is defined by

$$\dot{\mathbf{x}} = \mathbf{f}(\mathbf{x}, \mathbf{c}, t); \mathbf{x}(0) = \mathbf{x}_0. \quad (5.2)$$

Equations (5.1) and (5.2) are evaluated by numerical integration in the forward-time direction to yield a value of the objective function $f(\mathbf{c}) = J$.

Gradient vector $f_\mathbf{c}$ is evaluated conveniently in terms of Hamiltonian

$$H = F + \mathbf{p}'\mathbf{f} \quad (5.3)$$

where co-state vector p(*t*) is defined by

$$-\dot{\mathbf{p}} = H_{\mathbf{x}}, \mathbf{p}(t_f) = G_{\mathbf{x}_f}. \tag{5.4}$$

In particular, this gradient vector is given by

$$f_c = \int_0^{t_f} H_c dt + G_c, \tag{5.5}$$

and hence it can be evaluated by integrating equations (5.4) and (5.5) in the backward-time direction if state vector x(*t*) has been stored from the previous forward-time integration.

The scheme of forward-time and backward-time integrations outlined here is relatively easy to perform when a fixed time increment for integration is employed throughout. Numerical integration is performed using the Gill modification of the Runge-Kutta method [8], and straight-line interpolation is used for the midpoints required by this method of numerical integration. The scheme of forward-time and backward-time integrations described here is typical of two-point boundary-value problems that arise in optimization problems defined on a finite time interval. The particular two-point boundary-value problem for which LOAD3 is intended is summarized as follows:

Forward-time equations

$$\left.\begin{aligned}\dot{x}_0 &= F(\mathbf{x},\mathbf{c},t); x_0(0) = 0, J = x_0(t_f) + G(\mathbf{x}_f) \\ \dot{\mathbf{x}} &= \mathbf{f}(\mathbf{x},\mathbf{c},t); \mathbf{x}(0) = \mathbf{x}_o \\ \mathbf{c} &= \alpha\mathbf{d} + \mathbf{e}\end{aligned}\right\} \tag{5.6}$$

Backward-time equations

$$\left.\begin{aligned}-\dot{\mathbf{p}} &= H_{\mathbf{x}}, \mathbf{p}(t_f) = G_{\mathbf{x}_f} \\ -\dot{\mathbf{q}} &= H_{\mathbf{c}}, \mathbf{q}(t_f) = G_c \\ H &= F + \mathbf{p}'\mathbf{f} \\ J_c &= \mathbf{q}(0)\end{aligned}\right\} \tag{5.7}$$

Vectors **d** and **e** are computed in MIN1 or MIN3 as part of the method of successive approximations being employed. For example, steepest-descent results in

$$\mathbf{d} = -\mathbf{q}(0) \text{ and } \mathbf{e} = \mathbf{c}. \qquad (5.8)$$

LOAD3 and LOAD3S

Computation of the objective function and corresponding gradient vector for this class of parameter optimization problems is summarized as follows:

LOAD3 function: Compute

$$J = \int_0^{t_f} F(\mathbf{x}, \mathbf{c}, t)dt + G(\mathbf{x}_f, \mathbf{c})$$

and J_c where

$$\dot{\mathbf{x}} = \mathbf{f}(\mathbf{x}, \mathbf{c}, t); \mathbf{x}(0) = \mathbf{x}_o$$

for use with MIN1 or MIN3.

LOAD3 dummy argument list: MC,CV,J,DJV,MODE, where
    CV(MC) = **c**.
    J = *J*.
    DJV(MC) = J_c.
    MODE = 1 for program initialization and 2 otherwise.

LOAD3 input data list: NII,NP,NC,NX/DT/XV, where
    NII = number of integration increments.
    NP = number of integration increments omitted between output.
    $1 \leqslant NC \leqslant MC, 1 \leqslant NX \leqslant 7$.
    DT = length of integration increments.
    XV(NX) = $\mathbf{x}_o$.

LOAD3 output data list: None.

Subprograms called by LOAD3: LOAD3S for the computation of
    FV(NX) = $\mathbf{f}(\mathbf{y}, t)$, $F = F(\mathbf{y}, t)$, $G = G(\mathbf{y})$, DGV(NX+NC) = $G_\mathbf{y}$, DFV(NX+NC) = $F_\mathbf{y}$, and DFM(NX,NX+NC) = $\mathbf{f}_\mathbf{y}$, where

$$\mathbf{y} = \begin{bmatrix} \mathbf{x} \\ \mathbf{c} \end{bmatrix}.$$

Subprogram LOAD3 is listed in Table 5-1. This subprogram is used to spe-

Table 5-1
Listing of LOAD3

```
              SUBROUTINE LOAD3(MC,CV,F,DFV,MODE)
    C    DFM(NX,NX+NC),SXM(NX,NI+1),CV(NC),DV(NX+NC),DFV(NC),
    C   1DGV(NX+NC),DHV(NX+NC),DXV(NX),DYV(NX+NC),XV(NX),YV(NX+NC),
    C   2XOV(NX)
         REAL*8 CV(MC),DFV(MC),D,DF,DT,DTH,F,G,G1,G2,G3,G4,G5,
        1G6,G7,G8,T,DSQRT
         REAL*8 DFM(7,30),SXM(7,501),DV(30),DGV(30),DHV(30),DXV(7),
        1DYV(30),XV(7),YV(30),XOV(7)
         DATA MX/7/,MY/30/
         DATA DXV/7*0.D0/,DF/0.D0/,G/0.D0/,DGV/30*0.D0/,
        1DHV/30*0.D0/,DFM/210*0.D0/
    1000 FORMAT(//,'       T,XV,J')
    1001 FORMAT(4D18.10)
    1002 FORMAT(4I10)
    1003 FORMAT(4X,4D18.10)
    1004 FORMAT(//,'           NI,NP,NC,NX')
    1005 FORMAT('       DT')
         GO TO (1006,1007),MODE
    1006 READ (5,1002) NI,NP,NC,NX
         READ (5,1001) DT
         READ (5,1001) (XOV(I),I=1,NX)
         WRITE (6,1004)
         WRITE (6,1002) NI,NP,NC,NX
         WRITE (6,1005)
         WRITE (6,1003) DT
```

```
26      D=DSQRT(5D-1)
27      G1=DT*(1D0-D)
28      G2=DT*(1D0+D)
29      D=3D0*D
30      G6=D-2D0
31      G8=-(D+2D0)
32      G3=DT/6D0
33      G4=-DT/3D0
34      D=DSQRT(2D0)
35      G5=2D0-D
36      G7=2D0+D
37      DTH=5D-1*DT
38      N=NC+NX
39      CALL LOAD3S(MC,MX,MY,CV,XV,T,DXV,DF,G,DGV,DHV,DFM,1)
40      RETURN
41 1007 NIP=NP
42      NIC=0
43      T=0D0
44      DO 1008 I=1,NX
45      SXM(I,1)=XOV(I)
46 1008 XV(I)=XOV(I)
47      F=0D0
48      CALL LOAD3S(MC,MX,MY,CV,XV,T,DXV,DF,G,DGV,DHV,DFM,2)
49      WRITE (6,1000)
50 1009 CALL LOAD3S(MC,MX,MY,CV,XV,T,DXV,DF,G,DGV,DHV,DFM,3)
51      NIP=NIP+1
52      IF (NIP .LE. NP) GO TO 1010
```

(continued)

Table 5-1 Continued

```
53          WRITE (6,1003) T,(XV(I),I=1,NX),F
54          NIP=0
55    1010  DO 1011 I=1,NX
56          DV(I)=DXV(I)
57    1011  XV(I)=XV(I)+DTH*DXV(I)
58          D=DF
59          F=F+DTH*DF
60          T=T+DTH
61          CALL LOAD3S(MC,MX,MY,CV,XV,T,DXV,DF,G,DGV,DHV,DFM,3)
62          DO 1012 I=1,NX
63          XV(I)=XV(I)+G1*(DXV(I)-DV(I))
64    1012  DV(I)=G5*DXV(I)+G6*DV(I)
65          F=F+G1*(DF-D)
66          D=G5*DF+G6*D
67          CALL LOAD3S(MC,MX,MY,CV,XV,T,DXV,DF,G,DGV,DHV,DFM,3)
68          DO 1013 I=1,NX
69          XV(I)=XV(I)+G2*(DXV(I)-DV(I))
70    1013  DV(I)=G7*DXV(I)+G8*DV(I)
71          F=F+G2*(DF-D)
72          D=G7*DF+G8*D
73          T=T+DTH
74          CALL LOAD3S(MC,MX,MY,CV,XV,T,DXV,DF,G,DGV,DHV,DFM,3)
75          NIC=NIC+1
76          M=NIC+1
77          DO 1014 I=1,NX
78          XV(I)=XV(I)+G3*DXV(I)+G4*DV(I)
```

```
 79  1014 SXM(I,M)=XV(I)
 80       IF(NIC .LT. NI) GO TO 1009
 81       CALL LOAD3S(MC,MX,MY,CV,XV,T,DXV,DF,G,DGV,DHV,DFM,4)
 82       F=F+G
 83       WRITE(6,1003) T,(XV(I),I=1,NX),F
 84       DO 1015 I=1,N
 85  1015 YV(I)=DGV(I)
 86  1016 CALL LOAD3S(MC,MX,MY,CV,XV,T,DXV,DF,G,DGV,DHV,DFM,5)
 87       DO 1018 I=1,N
 88       D=DHV(I)
 89       DO 1017 J=1,NX
 90  1017 D=D+DFM(J,I)*YV(J)
 91  1018 DYV(I)=D
 92       DO 1019 I=1,N
 93       DV(I)=DYV(I)
 94  1019 YV(I)=YV(I)+DTH*DYV(I)
 95       T=T-DTH
 96       M=NIC+1
 97       DO 1020 I=1,NX
 98  1020 XV(I)=5D-1*(SXM(I,NIC)+SXM(I,M))
 99       CALL LOAD3S(MC,MX,MY,CV,XV,T,DXV,DF,G,DGV,DHV,DFM,5)
100       DO 1022 I=1,N
101       D=DHV(I)
102       DO 1021 J=1,NX
103  1021 D=D+DFM(J,I)*YV(J)
104  1022 DYV(I)=D
105       DO 1023 I=1,N
```

(continued)

Table 5-1 Continued

```
106            YV(I)=YV(I)+G1*(DYV(I)-DV(I))
107   1023  DV(I)=G5*DYV(I)+G6*DV(I)
108         DO 1025 I=1,N
109         D=DHV(I)
110         DO 1024 J=1,NX
111   1024  D=D+DFM(J,I)*YV(J)
112   1025  DYV(I)=D
113         DO 1026 I=1,N
114            YV(I)=YV(I)+G2*(DYV(I)-DV(I))
115   1026  DV(I)=G7*DYV(I)+G8*DV(I)
116         T=T-DTH
117         DO 1027 I=1,NX
118   1027  XV(I)=SXM(I,NIC)
119         CALL LOAD3S(MC,MX,MY,CV,XV,T,DXV,DF,G,DGV,DHV,DFM,5)
120         DO 1029 I=1,N
121         D=DHV(I)
122         DO 1028 J=1,NX
123   1028  D=D+DFM(J,I)*YV(J)
124   1029  DYV(I)=D
125         DO 1030 I=1,N
126   1030  YV(I)=YV(I)+G3*DYV(I)+G4*DV(I)
127         NIC=NIC-1
128         IF (NIC) 1031,1031,1016
129   1031  DO 1032 I=1,NC
130   1032  DFV(I)=YV(I+NX)
131         RETURN
132         END
```

cialize MIN1 or MIN3 to the class of parameter optimization problems defined by equations (5.1) and (5.2).

User problems are specified for subprogram LOAD3 by writing an appropriate subprogram LOAD3S. Specifications for writing this subprogram are summarized as follows:

LOAD3S function: Compute $\mathbf{f}(\mathbf{y}, t), F(\mathbf{y}, t), G(\mathbf{y}), G_\mathbf{y}, F_\mathbf{y}$ and $\mathbf{f}_\mathbf{y}$ for use with LOAD3, where

$$\mathbf{y} = \begin{bmatrix} \mathbf{x} \\ \mathbf{c} \end{bmatrix}.$$

LOAD3S dummy argument list: MC,MX,MY,CV,XV,T,FV,F,G,DGV,DFV,DFM,MODE, where
   CV(MC) = c.
   XV(MX) = x.
   T = t.
   FV(MX) = $\mathbf{f}(\mathbf{y}, t)$.
   F = $F(\mathbf{y}, t)$.
   G = $G(\mathbf{y})$.
   DGV(MY) = $G_\mathbf{y}$.
   DFV(MY) = $F_\mathbf{y}$.
   DFM(MX,MY) = $\mathbf{f}_\mathbf{y}$.
   MODE = 1 for program initialization, 2 for computations at $t = 0$, 3 for computation of FV and F, 4 for computation of G and DGV, and 5 for computation of DFV and DFM.

Example 7A.

An example of the use of subprogram LOAD3 is to compute the optimal parameter c_1 corresponding to the optimal gain control problem defined by

$$J = \int_0^{1/2} u_1^2 dt + \frac{e}{2} x_1^2\left(\frac{1}{2}\right),$$

$$\dot{x}_1 = u_1, \quad x_1(0) = 1,$$

and

$$u_1 = -c_1 x_1.$$

Calling program MIN1C and subprograms MIN1 and LOAD3S, which correspond to the above example of optimal gain control on a finite time interval, must be supplied by the user; they are listed in Table 5-2a. A listing of program input cards for this example is given in Table 5-2b. Finally, program printed output is listed in Table 5-2c for this example. Only the trial procedure of MIN1 is utilized for this example because there is only one parameter to adjust. Rapid convergence is achieved.

5.2 Optimal Control Functions with a Fixed Time Interval

Computer programs which are described in this section apply to a class of forcing-function optimization problems that arises in control. Namely, minimize

$$J = \int_0^{t_f} F(\mathbf{x},\mathbf{u},t)dt + G(\mathbf{x}_f) \tag{5.9}$$

with respect to control vector $\mathbf{u}(t)$ where state vector $\mathbf{x}(t)$ is defined by

$$\dot{\mathbf{x}} = \mathbf{f}(\mathbf{x},\mathbf{u},t),\ \mathbf{x}(0) = \mathbf{x}_o. \tag{5.10}$$

Final time t_f is fixed.

Description of the Basic Method

The programs described here, however, implement a parameter optimization technique for solving this class of optimization problems. The technique is taken directly from the derivation of the necessary conditions for optimality that are presented in chapter 6 of *Automated Design of Control Systems* [1]; specifically, perturbed control functions are expressed as vector $\mathbf{u}(t) + \alpha \mathbf{v}(t)$ where α is treated as an adjustable parameter and $\mathbf{v}(t)$ is treated as a direction vector for control vector $\mathbf{u}(t)$.

The mathematical form of direction vector $\mathbf{v}(t)$ can be deduced from the structural characteristics of two-point boundary-value problems [9]. In particular, the direction vector is of the form

$$\mathbf{v}(t) = \boldsymbol{\delta}(t) - \mathbf{K}(t)\mathbf{y}(t) \tag{5.11}$$

where vector $\mathbf{y}(t)$ is the corresponding direction vector in state vector $\mathbf{x}(t)$ which is defined by

$$\dot{\mathbf{y}} = \mathbf{f}_\mathbf{x}\mathbf{y} + \mathbf{f}_\mathbf{u}\mathbf{v},\ \mathbf{y}(0) = \mathbf{0}. \tag{5.12}$$

Table 5-2a
Listing of User-supplied Programs for Example 7A

```
 1      C    MIN1C  EXAMPLE 7A
 2                  CALL MIN1
 3                  STOP
 4                  END
 5                  SUBROUTINE MIN1S(MC,CV,F,DFV,MODE)
 6                  REAL*8 CV(MC),DFV(MC),F
 7                  CALL LOAD3(MC,CV,F,DFV,MODE)
 8                  RETURN
 9                  END
10                  SUBROUTINE LOAD3S(MC,MX,MY,CV,XV,T,FV,F,G,DGV,DFV,DFM,MODE)
11                  REAL*8 DFM(MX,MY),CV(MC),DFV(MY),DGV(MY),FV(MY),FV(MX),XV(MX),
12                 1F,G,T
13                  REAL*8 D,DD,DE,DF,E,DEXP
14                  GO TO (1000,1001,1002,1003,1004),MODE
15          1000    E=DEXP(1D0)
16                  D=E/2D0
17                  RETURN
18          1001    DD=CV(1)**2
19                  RETURN
20          1002    FV(1)=-CV(1)*XV(1)
21                  F=DD*XV(1)**2
22                  RETURN
23          1003    G=D*XV(1)**2
24                  DGV(1)=E*XV(1)
```

(*continued*)

Table 5-2a Continued

```
25          DFM(1,1)=-CV(1)
26          DE=2D0*CV(1)
27          DF=2D0*DD
28          RETURN
29     1004 DFV(1)=XV(1)*DF
30          DFV(2)=XV(1)**2*DE
31          DFM(1,2)=-XV(1)
32          RETURN
33          END
```

Table 5-2b
Listing of User-supplied Input Data for Example 7A

```
1   MIN1(LOAD3,LOAD3S,MIN1S)   EXAMPLE 7A
2   MERRIAM
3   UNIVERSITY OF ROCHESTER
4              1               5               1
5         0.              0.000001                     0.            5.
6         .1                  10.                      1            10.
7         0.                    1                      1             .5
8             20                1
9          .025
10         1.
```

Table 5-2c
Printed Output from MIN1 and LOAD3 for Example 7A

```
MIN1(LOAD3,LOAD3S,MIN1S)      EXAMPLE 7A
MERRIAM
UNIVERSITY OF ROCHESTER


NI,NT,NR,NC
  1       5           1            1
RATIO,DFMIN,FMIN,FMAX
  0.0000000000D 00  0.1000000000D-05  0.0000000000D 00  0.5000000000D 01
ALMIN,ALMAX,RMIN,RMAX
  0.1000000000D 00  0.1000000000D 02  0.5000000000D 00  0.1000000000D 02

NII,NP,NC,NX
  20       1           1            1
DT
  0.2500000000D-01

T,XV,J
  0.0000000000D 00  0.1000000000D 01  0.0000000000D 00
  0.5000000000D-01  0.1000000000D 01  0.0000000000D 00
  0.1000000000D 00  0.1000000000D 01  0.0000000000D 00
  0.1500000000D 00  0.1000000000D 01  0.0000000000D 00
```

(continued)

Table 5-2c Continued

```
0.2000000000D 00    0.1000000000D 01    0.0000000000D 00
0.2500000000D 00    0.1000000000D 01    0.0000000000D 00
0.3000000000D 00    0.1000000000D 01    0.0000000000D 00
0.3500000000D 00    0.1000000000D 01    0.0000000000D 00
0.4000000000D 00    0.1000000000D 01    0.0000000000D 00
0.4500000000D 00    0.1000000000D 01    0.0000000000D 00
0.5000000000D 00    0.1000000000D 01    0.1359140914D 01
```

ITERATION= 0   TRIAL= 0   F= 0.1359140914D 01
ALPHA= 0.0000000000D 00   DF= -0.1847264025D 01 |DFV|= 0.1359140914D 01
CV

0.0000000000D 00

T,XV,J

```
0.0000000000D 00    0.1000000000D 01    0.0000000000D 00
0.5000000000D-01    0.9048374229D 00    0.1813108032D 00
0.1000000000D 00    0.8187307620D 00    0.3297555352D 00
0.1500000000D 00    0.7408182327D 00    0.4512918038D 00
0.2000000000D 00    0.6703200606D 00    0.5507972855D 00
0.2500000000D 00    0.6065306762D 00    0.6322654844D 00
0.3000000000D 00    0.5488116540D 00    0.6989660050D 00
0.3500000000D 00    0.4965853227D 00    0.7535757730D 00
0.4000000000D 00    0.4493289836D 00    0.7982864700D 00
0.4500000000D 00    0.4065696796D 00    0.8348924930D 00
0.5000000000D 00    0.3678794611D 00    0.1048802711D 01
```

```
ITERATION= 0   TRIAL= 1    F=  0.1048802711D 01
ALPHA=  0.14715177765D 01  DF=  0.52180096835D 00    |DFV|=  0.38392610096D 00
CV
  0.20000000000D 01

T,XV,J
  0.00000000000D 00    0.10000000000D 01    0.00000000000D 00
  0.50000000000D-01    0.95182832600D 00    0.46422104190D-01
  0.10000000000D 00    0.90599771621D 00    0.88479470410D-01
  0.15000000000D 00    0.86233472560D 00    0.12658248370D 00
  0.20000000000D 00    0.82079461830D 00    0.16110294360D 00
  0.25000000000D 00    0.78125556750D 00    0.19237769180D 00
  0.30000000000D 00    0.74362117900D 00    0.22071189950D 00
  0.35000000000D 00    0.70779970190D 00    0.24638204450D 00
  0.40000000000D 00    0.67370308054D 00    0.26963860970D 00
  0.45000000000D 00    0.64125036530D 00    0.29070852660D 00
  0.50000000000D 00    0.61036026180D 00    0.81613126940D 00

ITERATION= 0   TRIAL= 2    F=  0.81613126940D 00
ALPHA=  0.72649700930D 00  DF= -0.11761798280D-01    |DFV|=  0.86538475580D-02
CV
  0.98741180930D 00

T,XV,J
  0.00000000000D 00    0.10000000000D 01    0.00000000000D 00
  0.50000000000D-01    0.95117899190D 00    0.47682382200D-01
```

(continued)

Table 5-2c Continued

```
0.100000000D  00    0.904741474 7D  00    0.9082261099D-01
0.150000000D  00    0.8605710838D  00    0.1298533652D  00
0.200000000D  00    0.8185571360D  00    0.1651661073D  00
0.250000000D  00    0.7785943514D  00    0.1971150097D  00
0.300000000D  00    0.7405825903D  00    0.2260205068D  00
0.350000000D  00    0.7044266017D  00    0.2521725088D  00
0.400000000D  00    0.6700357849D  00    0.2758333097D  00
0.450000000D  00    0.6373239624D  00    0.2972402176D  00
0.500000000D  00    0.6062091641D  00    0.8160780187D  00


ITERATION=  0    TRIAL=  3    F=  0.8160780187D  00
ALPHA=  0.7365390795D  00    DF=  0.1020585951D-02    IDFVI=  0.7509051788D-03
CV
 0.100106C398D  01

T,XV,J
 0.000000000D  00    0.1000000000D  01    0.0000000000D  00
 0.500000000D-01    0.9512323810D  00    0.4757812320D-01
 0.100000000D  00    0.9048430426D  00    0.9062885695D-01
 0.150000000D  00    0.8607160018D  00    0.1295830139D  00
 0.200000000D  00    0.8187409318D  00    0.1648304117D  00
 0.250000000D  00    0.7788128859D  00    0.1967237745D  00
 0.300000000D  00    0.7408320358D  00    0.2255822618D  00
 0.350000000D  00    0.7047034213D  00    0.2516946634D  00
```

```
 0.400000000D  00    0.670336713D 00    0.275322882D 00
 0.450000000D  00    0.637645987D 00    0.296701580D 00
 0.500000000D  00    0.606549511D 00    0.816077565D 00

ITERATION=  0    TRIAL=  4    F=  0.816077565D 00        |DFV|=  0.164783261D-04
ALPHA=  0.735713149D 00    DF= -0.223963672D-04
CV    0.999937842D 00

T,XV,J
 0.000000000D  00    0.100000000D 01    0.000000000D 00
 0.500000000D-01    0.951231487D 00    0.475798670D-01
 0.100000000D  00    0.904841342D 00    0.906320977D-01
 0.150000000D  00    0.860713576D 00    0.129587536D 00
 0.200000000D  00    0.818737855D 00    0.164836027D 00
 0.250000000D  00    0.778809228D 00    0.196730319D 00
 0.300000000D  00    0.740827860D 00    0.225589593D 00
 0.350000000D  00    0.704698788D 00    0.251702657D 00
 0.400000000D  00    0.670331676D 00    0.275330837D 00
 0.450000000D  00    0.637640597D 00    0.296710591D 00
 0.500000000D  00    0.606543814D 00    0.816077565D 00

ITERATION=  0    TRIAL=  5    F=  0.816077565D 00        |DFV|=  0.362763935D-05
ALPHA=  0.735729708D 00    DF= -0.493047306D-05
CV    0.999956627D 00

TRIAL LIMIT
```

Feedback gain matrix $\mathbf{K}(t)$ can be computed for the particular two-point boundary-value problem in question. Here, however, this matrix is taken to be a constant matrix $\mathbf{K}$ that is specified by the user via input data. Considerations in the selection of this matrix are discussed elsewhere [1].

Perturbation vector $\delta(t)$ is computed on each iteration, much in the same way that direction vector $\mathbf{d}$ is computed for the methods of successive approximation that are implemented in MIN1 and MIN3. Moreover, the trial procedure presented in chapter 2 for successive approximations can be applied to minimizing objective function $f(\alpha) = J$ in one dimension with respect to parameter α. In fact, programs MIN1 and MIN3 could be used directly to implement this method if gradient f_α could be computed from a backward-time integration. Forward-time integration of equation (5.12), which can be performed simultaneously with the integration of equation (5.10), is required to compute gradient f_α. Subprogram MIN4 is a modification of MIN1 that accommodates the integration of (5.12). User-supplied subprogram MIN4S is required to specialize subprogram MIN4 to particular design problems.

Forward-time and backward-time integrations, which are required to compute the objective function and its gradient with respect to parameter α, are implemented in subprogram FUN. These integrations are summarized as follows:

Forward-time equations

$$\left. \begin{aligned} \dot{x}_0 &= F(\mathbf{x}, \mathbf{u}, t), x_0(0) = 0, J = x_0(t_f) + G(\mathbf{x}_f) \\ \dot{\mathbf{x}} &= \mathbf{f}(\mathbf{x}, \mathbf{u}, t), \mathbf{x}(0) = \mathbf{x}_o \\ \dot{y}_0 &= \mathbf{y}'(F_\mathbf{x} - \mathbf{K}'F_\mathbf{u}) + \delta' F_\mathbf{u}, y_0(0) = 0, J_\alpha = y_0(t_f) + \mathbf{y}_f' G_{\mathbf{x}f} \\ \dot{\mathbf{y}} &= (\mathbf{f}_\mathbf{x} - \mathbf{f}_\mathbf{u}\mathbf{K})\mathbf{y} + \mathbf{f}_\mathbf{u}\delta, \mathbf{y}(0) = 0 \\ \mathbf{u} &= \alpha\delta + \mathbf{k} - \mathbf{K}\mathbf{x} \end{aligned} \right\} \quad (5.13)$$

Backward-time equations

$$\left. \begin{aligned} -\dot{\mathbf{p}} &= \mathbf{K}'H_\mathbf{u} + (F_\mathbf{x} - \mathbf{K}'F_\mathbf{u}) + \mathbf{A}'\mathbf{p}, \mathbf{p}(t_f) = G_{\mathbf{x}f} \\ -\dot{\mathbf{q}} &= -\mathbf{K}'H_\mathbf{u} + \mathbf{A}'\mathbf{q}, \mathbf{q}(t_f) = 0 \\ H &= F + \mathbf{p}'\mathbf{f}, \mathbf{A} = H_{\mathbf{px}} - H_{\mathbf{pu}}\mathbf{K} \\ \delta &= -\mathbf{R}^{-1}(H_\mathbf{u} + H_{\mathbf{up}}\mathbf{q}), \mathbf{k} = \mathbf{u} + \mathbf{K}\mathbf{x} \end{aligned} \right\} \quad (5.14)$$

Matrix **R** is positive definite and is specified in subprogram FUN as a constant matrix via input data.

MIN4, MIN4S, and FUN

The method of successive approximations outlined previously in this section is implemented by subprogram MIN4, which is summarized as follows:

MIN4 function: Modified steepest-descent method of successive approximations for minimizing

$$J = \int_0^{t_f} F(\mathbf{x}, \mathbf{u}, t)dt + G(\mathbf{x}_f)$$

with respect to $\mathbf{u}(t)$, where

$$\dot{\mathbf{x}} = \mathbf{f}(\mathbf{x}, \mathbf{u}, t), \mathbf{x}(0) = \mathbf{x}_o.$$

MIN4 dummy argument list: None.

MIN4 input data list: C1/C2/C3/NI,NT/RATIO,DFMIN,FMIN,FMAX/ALMIN, ALMAX,RMIN,RMAX, where
   NI = maximum number of iterations.
   NT = maximum number of trials per iteration.
   RATIO = minimum value of $|s(\alpha)/s(0)|$ requiring additional trials.
   DFMIN = minimum value of $|f_c|$ requiring additional trials and iterations.
   FMIN = lower bound on $f(\mathbf{c})$.
   FMAX = upper bound on $f(\mathbf{c})$.
   ALMIN = minimum initial value of α.
   ALMAX = maximum initial value of α.
   RMIN = minimum fraction of $(\alpha - \beta)$ used in interval reduction (<1).
   RMAX = maximum fraction of $(\alpha - \beta)$ used in extrapolation (>1).

MIN4 output data list: None.

Subprograms called by MIN4: FUN and MIN4S for the computation of
   FV(NX) = $\mathbf{f}(\mathbf{y}, t)$, F = $F(\mathbf{y}, t)$, G = $G(\mathbf{x})$,
   DGV(NX) = $G_\mathbf{x}$, DFM(NX,NX+NU) = $\mathbf{f}_\mathbf{y}$, and DFV(NX+NU) = $F_\mathbf{y}$, where

$$\mathbf{y} = \begin{bmatrix} \mathbf{x} \\ \mathbf{u} \end{bmatrix}$$

Subprogram MIN4 is listed in Table 5-3. User problems are specified for sub-

Table 5-3
Listing of MIN4

```
1            SUBROUTINE MIN4
2            REAL*8 A,ALMIN,ALMAX,B,C,DC,DF,DFA,DFB,DFMIN,X,
3           1D1,D2,D3,F,FA,FB,FMIN,FMAX,RATIO,RMIN,RMAX,SDF,D4
4            REAL*8 DABS,DMIN1,DMAX1,DSQRT
5            LOGICAL*1 COM(65)
6       1000 FORMAT(65A1)
7       1001 FORMAT(4X,65A1)
8       1002 FORMAT(4D18.10)
9       1003 FORMAT(4X,4D18.10)
10      1004 FORMAT(4I10)
11      1005 FORMAT(' ')
12      1006 FORMAT(//,'           NI,NT')
13      1007 FORMAT('     RATIO,DHMIN,JMIN,JMAX')
14      1008 FORMAT('     ALMIN,ALMAX,RMIN,RMAX')
15      1009 FORMAT(//,'     ITERATION=',I2,'  TRIAL=',I2,'   J=',D18.10,
16      1010 FORMAT('     ALPHA=',D18.10,'  DJ=',D18.10,
17           1'  (DHV(=',D18.10)
18      1011 FORMAT(//,'    RESET KV,KM')
19      1012 FORMAT(//,'    DESIGN COMPLETE')
20      1013 FORMAT(//,'    ITERATION LIMIT')
21      1014 FORMAT(//,'    TRIAL LIMIT')
22           WRITE (6,1005)
23           DO 1015 I=1,3
24           READ (5,1000) COM
```

```
25  1015 WRITE (6,1001) CCM
26       READ (5,1004) NI,NT
27       READ (5,1002) RATIO,DFMIN,FMIN,FMAX
28       READ (5,1002) ALMIN,ALMAX,RMIN,RMAX
29       WRITE (6,1006)
30       WRITE (6,1004) NI,NT
31       WRITE (6,1007)
32       WRITE (6,1003) RATIO,DFMIN,FMIN,FMAX
33       WRITE (6,1008)
34       WRITE (6,1003) ALMIN,ALMAX,RMIN,RMAX
35       NIC=0
36       NTC=0
37       C=0.D0
38       X=1.D0-1.D0/RMAX
39       CALL FUN(DFMIN,C,F,DF,D3,1)
40       CALL FUN(DFMIN,C,F,DF,D3,2)
41       SDF=RATIO*DABS(DF)
42       WRITE (6,1009) NIC,NTC,F
43       WRITE (6,1010) C,DF,D3
44       IF (FMAX .GT. F) GO TO 1016
45       WRITE (6,1011)
46       RETURN
47  1016 IF (DFMIN .LT. D3) GO TO 1017
48       WRITE (6,1012)
49       WRITE (6,1005)
50       CALL FUN(DFMIN,C,F,DF,D3,5)
51       RETURN
52  1017 DC=2.D0*(FMIN-F)/DF
```

(continued)

Table 5-3 Continued

```
53   1018 DC=DMAX1(ALMIN,DMIN1(DC,ALMAX))
54   1019 NTC=NTC+1
55        A=C
56        FA=F
57        DFA=DF
58        C=A+DC
59        CALL FUN(DFMIN,C,F,DF,D3,3)
60        IF (SDF .GE. DABS(DF)) GO TO 1033
61        WRITE (6,1009) NIC, NTC, F
62        WRITE (6,1010) C,DF,D3
63        IF (FMAX .GT. F) GO TO 1022
64   1020 IF (NT .GT. NTC) GO TO 1021
65        GO TO 1026
66   1021 DC=RMIN*DC
67        C=A
68        F=FA
69        DF=DFA
70        GO TO 1019
71   1022 IF (DFMIN .LT. D3) GO TO 1023
72        WRITE (6,1012)
73        GO TO 1034
74   1023 IF (DF) 1024,1024,1027
75   1024 IF (NT .LE. NTC) GO TO 1026
76        IF (X*DFA .LE. DF) GO TO 1025
77        DC=DC*RMAX
78        GO TO 1019
```

```
 79  1025 DC=DC*DFA/(DFA-DF)
 80       GO TO 1019
 81  1026 WRITE (6,1014)
 82       GO TO 1034
 83  1027 B=C
 84       FB=F
 85       IFB=DF
 86  1028 D1=DFA+DFB+3.D0*(FA-FB)/DC
 87       D2=2.D0*(DFA+D1)
 88       D4=DFA+DFB+2.D0*D1
 89       IF (1.D-5*DABS(D2) .LE. DABS(D4)) GO TO 1029
 90       C=B-(DFA+2.D0*D1)*DC/D2
 91       GO TO 1030
 92  1029 D2=DSQRT(D1**2-DFA*DFB)
 93       C=B-(DFB+D1-D2)*DC/D4
 94  1030 CALL FUN (DFMIN,C,F,DF,D3,3)
 95       IF (SDF .GE. DABS(DF)) GO TO 1033
 96       NTC=NTC+1
 97       WRITE (6,1009) NIC,NTC,F
 98       WRITE (6,1010) C,DF,D3
 99       IF (FMAX .LE. F) GO TO 1020
100       IF (NT .LE. NTC) GO TO 1026
101       IF (DF) 1032,1031,1031
102  1031 DC=C-A
103       GO TO 1027
104  1032 A=C
105       FA=F
```

(continued)

Table 5-3 Continued

```
106          DFA=DF
107          DC=B-C
108          GO TO 1028
109  1033    NIC=NIC+1
110          CALL FUN(DFMIN,C,F,DF,D3,4)
111          DC=C
112          C=0.D0
113          NTC=0
114          SDF=RATIO*DABS(DF)
115          WRITE (6,1009) NIC,NTC,F
116          WRITE (6,1010) C,DF,D3
117          IF (NI .GT. NIC) GO TO 1018
118          WRITE (6,1013)
119  1034    WRITE (6,1005)
120          CALL FUN(DFMIN,C,F,DF,D3,5)
121          RETURN
122          END
```

program MIN4 by writing an appropriate subprogram MIN4S. Specifications for writing this subprogram are given as follows:

MIN4S function: Compute $\mathbf{f}(\mathbf{y}, t)$, $F(\mathbf{y}, t)$, $G(\mathbf{x})$, $G_\mathbf{x}$, $\mathbf{f_y}$, and $F_\mathbf{y}$ for use with MIN4, where

$$\mathbf{y} = \begin{bmatrix} \mathbf{x} \\ \mathbf{u} \end{bmatrix}.$$

MIN4S dummy argument list: MX,MY,YV,T,FV,F,G,DGV,DFM,DFV,MODE, where
  YV(MY) = $\mathbf{y}$.
  T = t.
  FV(MX) = $\mathbf{f}(\mathbf{y}, t)$.
  F = $F(\mathbf{y}, t)$.
  G = $G(\mathbf{x})$.
  DGV(MX) = $G_\mathbf{x}$.
  DFM(MX,MY) = $\mathbf{f_y}$.
  DFV(MY) = $F_\mathbf{y}$.
  MODE = 1 program initialization, 2 for computation of FV and F, 3 for computation of G, 4 for computation of DGV, and 5 for computation of DFM and DFV.

Subprogram MIN4S is actually called by a subprogram FUN which implements forward-time and backward-time integrations and in turn is called by MIN4. Subprogram FUN is summarized as follows:

FUN function: Compute

$$J = \int_0^{t_f} F(\mathbf{x}, \mathbf{u}, t) dt + G(\mathbf{x}_f),$$

J_α, and $\|H_\mathbf{u}\|$, where

$$\dot{\mathbf{x}} = \mathbf{f}(\mathbf{x}, \mathbf{u}, t), \mathbf{x}(0) = \mathbf{x}_o$$

and

$$\mathbf{u} = \mathbf{k} - K\mathbf{x}.$$

FUN dummy argument list: DHMIN,ALPHA,J,DJ,DHN,MODE, where
  DHMIN = minimum value of $|H_{u_i}|$ requiring additional trials and iterations.
  ALPHA = α.
  J = J.

DJ = J_α.
DHN = $\|H_u\|$.
MODE = 1 for program initialization, 2 for forward- and backward-time integrations required, 3 for the forward-time integration required, 4 for the backward-time integration required, and 5 for I/O output of k.

FUN input data list: NII,NP,NX,NU,MTAPE/DT/RIM/KM/DV, where
NII = number of integration increments.
NP = number of integration increments omitted between output. $1 \leq NX \leq 7, 1 \leq NU \leq 3$.
MTAPE = 1 for constant k input on cards or 2 for time varying k input on I/O unit 3.
DT = length of integration increments.
RIM(NU,NU) = $\mathbf{R}^{-1}$.
KM(NU,NX) = $\mathbf{K}$.
KV(NU) = k (either on cards or on I/O unit 3).

FUN output data list: KV, where
KV(NU) = k (on I/O unit 4).

FUN subprograms called: MIN4S.

Subprogram FUN is listed in Table 5–4.

Example 7B

An example of the use of subprograms MIN4 and FUN is provided by the problem of computing optimal control function $u_1(t)$ corresponding to minimizing

$$ J = \int_0^{1/2} u_1^2 dt + \frac{e}{2} x_1^2 \left(\frac{1}{2}\right) $$

where

$$ \dot{x}_1 = u_1, \quad x_1(0) = 1. $$

Calling program MIN4C and the subprogram MIN4S, which correspond to the above example, must be supplied by the user; they are listed in Table 5–5a. A listing of program input cards for this example is given in Table 5–5b. Finally, program printed output is listed in Table 5–5c for this example. Iterations are initialized with the solution found for Example 7A by selecting feedback gain matrix $\mathbf{K} = [1]$. A comparison of the first and last iterations, therefore, illustrates differences between solutions to parameter and forcing-function optimization problems on a finite time interval.

Table 5-4
Listing of FUN

```
 1            SUBROUTINE FUN (DHMIN,C,X,XA,DH,MODE)
 2   C        REAL*8 AM(NX,NX),DFM(NX,NX+NU),KM(NU,NX),RIM(NU,NU),
 3   C       1SYM(NX+NU,NTAB),SZM(2*NU,NTAB),AV(NX),DV(NX),DFV(NX+NU),
 4   C       2DGV(NX),DXAV(NX),DAV(NX),DPV(NX),FV(NX),HV(NU),PV(NX),
 5   C       3XAV(NX),XOV(NX),YV(NX+NU),ZV(2*NU)
 6            REAL*8 AM(7,7),DFM(7,10),KM(3,7),RIM(3,3),SYM(10,401),
 7           1SZM(6,401),AV(7),DV(7),DAV(7),DFV(10),DGV(7),DPV(7),
 8           2DXAV(7),FV(7),HV(3),PV(7),XAV(7),XOV(7),YV(10),ZV(6)
 9            REAL*8 C,D,DA,DD,DH,DT,DTH,DXA,DHMIN,F,G,G1,G2,G3,G4,G5,
10           1G6,G7,G8,X,XA,DABS,DSQRT,T
11            DATA MX/7/,MY/10/
12            DATA FV/7*0.D0/,F/0.D0/,G/0.D0/,DGV/7*0.D0/,DFM/70*0.D0/,
13           1DFV/10*0.D0/
14       1000 FORMAT(' RIM')
15       1001 FORMAT(4D18.10)
16       1002 FORMAT(5I10)
17       1003 FORMAT(4X,4D18.10)
18       1004 FORMAT(///,'     NEI,NP,NX,NU,MTAPE')
19       1005 FORMAT('   DT')
20       1006 FORMAT('   KV')
21       1007 FORMAT('   KM')
22       1008 FORMAT(///,'   T,YV,J')
23       1009 FORMAT(///,'   T,KV,DELV')
24            GO TO (1010,1017,1017,1063,1096),MODE
```

(continued)

Table 5-4 Continued

```
25   1010 READ (5,1002) NI,NP,NX,NU,NPRIME
26        READ (5,1001) DT
27        READ (5,1001) ((RIM(I,J),I=1,J),J=1,NU)
28        READ (5,1001) ((KM(I,J),I=1,NU),J=1,NX)
29        READ (5,1001) (XOV(I),I=1,NX)
30        WRITE (6,1004)
31        WRITE (6,1002) NI,NP,NX,NU,NPRIME
32        WRITE (6,1005)
33        WRITE (6,1003) DT
34        WRITE (6,1000)
35        WRITE (6,1003) ((RIM(I,J),I=1,J),J=1,NU)
36        WRITE (6,1007)
37        WRITE (6,1003) ((KM(I,J),I=1,NU),J=1,NX)
38        NTAB=NI+1
39        GO TO (1011,1013),NPRIME
40   1011 READ (5,1001) (ZV(I),I=1,NU)
41        WRITE (6,1006)
42        WRITE (6,1003) (ZV(I),I=1,NU)
43        DO 1012 I=1,NU
44        DO 1012 J=1,NTAB
45   1012 SZM(I,J)=ZV(I)
46        GO TO 1014
47   1013 READ (3) ((SZM(I,J),I=1,NU),J=1,NTAB)
48        REWIND 3
49   1014 DO 1016 I=1,NU
50        DO 1015 J=1,NU
```

```
51      1015  RIM(J,I)=RIM(I,J)
52            DO 1016 J=1,NTAB
53      1016  SZM(I+NU,J)=0D0
54            D=DSQRT(5D-1)
55            G1=DT*(1D0-D)
56            G2=DT*(1D0+D)
57            D=3D0*D
58            G6=D-2D0
59            G8=-(D+2D0)
60            G3=DT/6D0
61            G4=-DT/3D0
62            D=DSQRT(2D0)
63            G5=2D0-D
64            G7=2D0+D
65            DTH=5D-1*DT
66            NY=NX+NU
67            NZ=2*NU
68            CALL MIN4S(MX,MY,YV,T,FV,F,G,DGV,DFM,DFV,1)
69            RETURN
70      1017  NIP=NP
71            NIC=0
72            L=0
73            T=0D0
74            DO 1018 I=1,NX
75      1018  YV(I)=XOV(I)
76            X=0D0
77            IC=1
```

Table 5-4 Continued

```
78              IF (MODE .EQ. 2) GO TO 1020
79              IC=0
80              DO 1019 I=1,NX
81        1019  XAV(I)=0D0
82              XA=0D0
83        1020  DO 1021 I=1,NZ
84        1021  ZV(I)=SZM(I,1)
85              NGILL=1
86              WRITE (6,1008)
87        1022  DO 1024 I=1,NU
88              D=ZV(I)+C*ZV(I+NU)
89              DO 1023 J=1,NX
90        1023  D=D-KM(I,J)*YV(J)
91        1024  YV(I+NX)=D
92              CALL MIN4S(MX,MY,YV,T,FV,F,G,DGV,DFM,DFV,2)
93              IF (IC) 1025,1025,1033
94        1025  CALL MIN4S(MX,MY,YV,T,FV,F,G,DGV,DFM,DFV,5)
95              DO 1028 I=1,NX
96              D=DFV(I)
97              DO 1026 J=1,NU
98        1026  D=D-KM(J,I)*DFV(J+NX)
99              AV(I)=D
100       1028  DO 1028 J=1,NX
101             D=DFM(I,J)
102             DO 1027 K=1,NU
103       1027  D=D-DFM(I,K+NX)*KM(K,J)
```

```
104       1028  AM(I,J)=D
105             DXA=0D0
106             DO 1031 I=1,NX
107             DXA=DXA+XAV(I)*AV(I)
108             D=0D0
109             DO 1029 J=1,NX
110       1029  D=D+AM(I,J)*XAV(J)
111             DO 1030 J=1,NU
112       1030  D=D+DFM(I,J+NX)*ZV(J+NU)
113       1031  DXAV(I)=D
114             DO 1032 I=1,NU
115       1032  DXA=DXA+ZV(I+NU)*DFV(I+NX)
116       1033  GO TO (1034,1043,1048,1054),NGILL
117       1034  NIC=NIC+1
118             DO 1035 I=1,NY
119       1035  SYM(I,NIC)=YV(I)
120             NIP=NIP+1
121             IF (NIP .LE. NP) GO TO 1037
122             IF (L) 1036,1036,1059
123       1036  WRITE (6,1003) T,(YV(I),I=1,NY),X
124             NIP=0
125       1037  IF (IC) 1038,1038,1040
126       1038  DO 1039 I=1,NX
127             DAV(I)=DXAV(I)
128       1039  XAV(I)=XAV(I)+DTH*DXAV(I)
129             DA=DXA
130             XA=XA+DTH*DXA
```

(continued)

Table 5-4 Continued

```
1040    DO 1041 I=1,NX
        DV(I)=FV(I)
1041    YV(I)=YV(I)+DTH*FV(I)
        DD=F
        X=X+DTH*F
        T=T+DTH
        M=NIC+1
        DO 1042 I=1,NZ
1042    ZV(I)=5D-1*(SZM(I,NIC)+SZM(I,M))
        NGILL=2
        GO TO 1022
1043    IF (IC)   1044,1044,1046
1044    DO 1045 I=1,NX
        XAV(I)=XAV(I)+G1*(DXAV(I)-DAV(I))
1045    DAV(I)=G5*DXAV(I)+G6*DAV(I)
        XA=XA+G1*(DXA-DA)
        DA=G5*DXA+G6*DA
1046    DO 1047 I=1,NX
        YV(I)=YV(I)+G1*(FV(I)-DV(I))
1047    DV(I)=G5*FV(I)+G6*DV(I)
        X=X+G1*(F-DD)
        DD=G5*F+G6*DD
        NGILL=3
        GO TO 1022
1048    IF (IC)   1049,1049,1051
1049    DO 1050 I=1,NX
```

```
157        XAV(I)=XAV(I)+G2*(DXAV(I)-DAV(I))
158 1050   DAV(I)=G7*DXAV(I)+G8*DAV(I)
159        XA=XA+G2*(DXA-DA)
160        DA=G7*DXA+G8*DA
161 1051   DO 1052 I=1,NX
162        YV(I)=YV(I)+G2*(FV(I)-DV(I))
163 1052   DV(I)=G7*FV(I)+G8*DV(I)
164        X=X+G2*(F-DD)
165        DD=G7*F+G8*DD
166        T=T+DTH
167        DO 1053 I=1,NZ
168 1053   ZV(I)=SZM(I,M)
169        NGILL=4
170        GO TO 1022
171 1054   IF (IC)  1055,1055,1057
172 1055   DO 1056 I=1,NX
173 1056   XAV(I)=XAV(I)+G3*DXAV(I)+G4*DAV(I)
174        XA=XA+G3*DXA+G4*DA
175 1057   DO 1058 I=1,NX
176 1058   YV(I)=YV(I)+G3*FV(I)+G4*DV(I)
177        X=X+G3*F+G4*DD
178        NGILL=1
179        IF (NIC .LT. NI) GO TO 1022
180        L=1
181        NIP=NP
182        GO TO 1022
183 1059   CALL MIN4S(MX,MY,YV,T,FV,F,G,DGV,DFM,DFV,3)
```

Table 5-4 Continued

```
184            X=X+G
185            WRITE (6,1003) T,(YV(I),I=1,NY),X
186            IF (IC) 1060,1060,1062
187      1060  CALL MIN4S(MX,MY,YV,T,FV,F,G,DGV,DFM,DFV,4)
188            DO 1061 I=1,NX
189      1061  XA=XA+XAV(I)*DGV(I)
190            RETURN
191      1062  CALL MIN4S(MX,MY,YV,T,FV,F,G,DGV,DFM,DFV,4)
192      1063  NIP=NP
193            L=0
194            DO 1064 I=1,NX
195            PV(I)=DGV(I)
196      1064  XAV(I)=0DQ
197            XA=0D0
198            DH=0D0
199            NGILL=1
200            WRITE (6,1009)
201      1065  CALL MIN4S(MX,MY,YV,T,FV,F,G,DGV,DFM,DFV,5)
202      1066  DO 1071 I=1,NU
203            D=DFV(I+NX)
204            DO 1067 J=1,NX
205      1067  D=D+DFM(J,I+NX)*PV(J)
206            IF (DABS(D) .GT. DHMIN) GO TO 1068
207            D=0D0
208      1068  IF (DABS(D) .LE. DH) GO TO 1069
```

```
209         DH=DABS(D)
210  1069   HV(I)=D
211         DO 1070  J=1,NX
212  1070   D=D+DFM(J,I+NX)*XAV(J)
213  1071   ZV(I)=D
214         DXA=0D0
215         DO 1073  I=1,NU
216         D=0D0
217         DO 1072  J=1,NU
218  1072   D=D-RIM(I,J)*ZV(J)
219         DXA=DXA+D*ZV(I)
220  1073   ZV(I+NU)=D
221         DO 1075  I=1,NU
222         D=YV(I+NX)
223         DO 1074  J=1,NX
224  1074   D=D+KM(I,J)*YV(J)
225  1075   ZV(I)=D
226         DO 1078  I=1,NX
227         D=DFV(I)
228         DO 1076  J=1,NU
229  1076   D=D-KM(J,I)*DFV(J+NX)
230         AV(I)=D
231         DO 1078  J=1,NX
232         D=DFM(I,J)
233         DO 1077  K=1,NU
234  1077   D=D-DFM(I,K+NX)*KM(K,J)
235  1078   AM(I,J)=D
```

(continued)

Table 5-4 Continued

```
236              DO 1081 I=1,NX
237              DD=0D0
238              DO 1079 J=1,NU
239         1079 DD=DD-KM(J,I)*HV(J)
240              D=AV(I)-DD
241              DO 1080 J=1,NX
242              D=D+AM(J,I)*PV(J)
243         1080 DD=DD+AM(J,I)*XAV(J)
244              DPV(I)=D
245         1081 DXAV(I)=DD
246              GO TO (1082,1088,1090,1093),NGILL
247         1082 DO 1083 I=1,NZ
248         1083 SZM(I,NIC)=ZV(I)
249              NIP=NIP+1
250              IF (NIP .LE. NP) GO TO 1085
251              WRITE (6,1003) T,(ZV(I),I=1,NZ)
252              IF (L) 1084,1084,1095
253         1084 NIP=0
254         1085 DO 1086 I=1,NX
255              DV(I)=DPV(I)
256              DAV(I)=DXAV(I)
257              PV(I)=PV(I)+DTH*DPV(I)
258         1086 XAV(I)=XAV(I)+DTH*DXAV(I)
259              DA=DXA
260              XA=XA+DTH*DXA
```

(continued)

```
261              T=T-DTH
262              M=NIC-1
263              DO 1087 I=1,NY
264         1087 YV(I)=5D-1*(SYM(I,M)+SYM(I,NIC))
265              NIC=M
266              NGILL=2
267              GO TO 1065
268         1088 DO 1089 I=1,NX
269              PV(I)=PV(I)+G1*(DPV(I)-DV(I))
270              XAV(I)=XAV(I)+G1*(DXAV(I)-DAV(I))
271              DV(I)=G5*DPV(I)+G6*DV(I)
272         1089 DAV(I)=G5*DXAV(I)+G6*DAV(I)
273              XA=XA+G1*(DXA-DA)
274              DA=G5*DXA+G6*DA
275              NGILL=3
276              GO TO 1066
277         1090 DO 1091 I=1,NX
278              PV(I)=PV(I)+G2*(DPV(I)-DV(I))
279              XAV(I)=XAV(I)+G2*(DXAV(I)-DAV(I))
280              DV(I)=G7*DPV(I)+G8*DV(I)
281         1091 DAV(I)=G7*DXAV(I)+G8*DAV(I)
282              XA=XA+G2*(DXA-DA)
283              DA=G7*DXA+G8*DA
284              T=T-DTH
285              DO 1092 I=1,NY
286         1092 YV(I)=SYM(I,NIC)
287              NGILL=4
```

Table 5–4 Continued

```
288            GO TO 1065
289  1093  DO 1094 I=1,NX
290        PV(I)=PV(I)+G3*DPV(I)+G4*DV(I)
291  1094  XAV(I)=XAV(I)+G3*DXAV(I)+G4*DAV(I)
292        XA=XA+G3*DXA+G4*DA
293        NGILL=1
294        IF (NIC .GT. 1) GO TO 1066
295        L=1
296        NIP=NP
297        GO TO 1066
298  1095  RETURN
299  1096  WRITE (4) ((SZM(I,J),I=1,NU),J=1,NTAB)
300        END FILE 4
301        REWIND 4
302        RETURN
303        END
```

Table 5-5a
Listing of User-supplied Programs for Example 7B

```
1     C     MIN4C    EXAMPLE 7B
2           CALL MIN4
3           STOP
4           END
5           SUBROUTINE MIN4S(MX,MY,YV,T,FV,F,G,DGV,DFM,DFV,MODE)
6           REAL*8 DFM(MX,MY),DFV(MY),DGV(MX),FV(MX),YV(MY),F,G,T
7           REAL*8 D,E,DEXP
8           GO TO (1000,1001,1002,1003,1004),MODE
9     1000  E=DEXP(1D0)
10          D=E/2D0
11          DFM(1,2)=1D0
12          RETURN
13    1001  F=YV(2)**2
14          FV(1)=YV(2)
15          RETURN
16    1002  G=D*YV(1)**2
17          RETURN
18    1003  DGV(1)=E*YV(1)
19          RETURN
20    1004  DFV(2)=2D0*YV(2)
21          RETURN
22          END
```

Table 5-5b
Listing of User-supplied Input Data for Example 7B

```
1   MIN4(FUN,MIN4S)    EXAMPLE 7B
2   MERRIAM
3   UNIVERSITY OF ROCHESTER
4          2                    4        0.        5.
5    .01                                 10.       10.
6    .1
7           20           1            1
8    .025                          0.       .5       1
9    .5                         1                    1
10   1.
11   1.
12   0.
```

Table 5-5c
Printed Output from MIN4 and FUN for Example 7B

```
MIN4(FUN,MIN4S)    EXAMPLE 7B
MERRIAM
UNIVERSITY OF ROCHESTER

NI,NT
   2          4
RATIO,DHMIN,JMIN,JMAX
 0.1000000000D-01  0.0000000000D 00  0.0000000000D 00  0.5000000000D 01
ALMIN,ALMAX,RMIN,RMAX
 0.1000000000D 00  0.1000000000D 02  0.5000000000D 00  0.1000000000D 02

NII,NP,NX,NU,MTAPE
  20      1       1       1       1
DT
 0.2500000000D-01
RIM
 0.5000000000D 00
KM
 0.1000000000D 01
KV
 0.0000000000D 00
```

(continued)

Table 5-5c Continued

T, YV, J

0.000000000D 00	0.000000000D 00	01	0.000000000D 00
0.500000000D-01	0.100000000D	-0.100000000D 01	0.475812910 6D-01
0.100000000D 00	0.951229424 7D 00	-0.951229424 7D 00	0.906346236 3D-01
0.150000000D 00	0.904837418 3D 00	-0.904837418 3D 00	0.129590889 9D 00
0.200000000D 00	0.860707976 9D 00	-0.860707976 9D 00	0.164839977 3D 00
0.250000000D 00	0.818730753 6D 00	-0.818730753 6D 00	0.196734670 6D 00
0.300000000D 00	0.778800783 7D 00	-0.778800783 7D 00	0.225594182 5D 00
0.350000000D 00	0.740818221 4D 00	-0.740818221 4D 00	0.251707348 7D 00
0.400000000D 00	0.704688090 5D 00	-0.704688090 5D 00	0.275335518 7D 00
0.450000000D 00	0.670320046 9D 00	-0.670320046 9D 00	0.296715171 0D 00
0.500000000D 00	0.637628152 6D 00	-0.637628152 6D 00	0.816060282 0D 00
	0.606530660 7D 00	-0.606530660 7D 00	

T, KV, DELV

0.500000000D 00	0.000000000D 00	-0.217829976 0D 00
0.450000000D 00	0.000000000D 00	-0.176868692 3D 00
0.400000000D 00	0.000000000D 00	-0.136349672 5D 00
0.350000000D 00	0.000000000D 00	-0.961715979 3D-01
0.300000000D 00	0.000000000D 00	-0.562340024 2D-01
0.250000000D 00	0.000000000D 00	-0.164370211 9D-01
0.200000000D 00	0.000000000D 00	0.233188589 1D-01
0.150000000D 00	0.000000000D 00	0.631330483 1D-01
0.100000000D 00	0.000000000D 00	0.103105103 2D 00
0.500000000D-01	0.000000000D 00	0.143334974 6D 00
-0.473579508 9D-15	0.000000000D 00	0.183923258 0D 00

ITERATION= 0   TRIAL= 0   J= 0.8160602820D 00
ALPHA= 0.000000000D 00   DJ= -0.1361525518D-01 |DHV|= 0.4356599520D 00

T,YV,J
0.0000000000D 00   0.1000000000D 01   0.8392325804D 00   0.0000000000D 00
0.5000000000D-01   0.1030937343D 01   0.4024124026D 00   0.1993743255D-01
0.1000000000D 00   0.1040661921D 01   -0.9610889317D-02   0.2253612644D-01
0.1500000000D 00   0.1030358335D 01   -0.3990278517D 00   0.2529125959D-01
0.2000000000D 00   0.1001104515D 01   -0.7679159256D 00   0.4297397140D-01
0.2500000000D 00   0.9538771792D 00   -0.1118247391D 01   0.8809377836D-01
0.3000000000D 00   0.8895569006D 00   -0.1451896925D 01   0.1712995837D 00
0.3500000000D 00   0.8089328045D 00   -0.1770648784D 01   0.3017278253D 00
0.4000000000D 00   0.7127069182D 00   -0.2076203644D 01   0.4873052644D 00
0.4500000000D 00   0.6014981861D 00   -0.2370185109D 01   0.7350130110D 00
0.5000000000D 00   0.4758461663D 00   -0.2654145926D 01   0.1358867285D 01

ITERATION= 0   TRIAL= 1   J= 0.1358867285D 01
ALPHA= 0.1000000000D 02   DJ= 0.1221777782D 00 |DHV|= 0.4356599520D 00

T,YV,J
0.0000000000D 00   0.1000000000D 01   -0.8155856823D 00   0.0000000000D 00
0.5000000000D-01   0.9592214979D 00   -0.8155038309D 00   0.3325772478D-01
0.1000000000D 00   0.9184561327D 00   -0.8150757475D 00   0.6649402555D-01
0.1500000000D 00   0.8777183079D 00   -0.8144166996D 00   0.9968543471D-01

(continued)

Table 5-5c Continued

```
 0.200000000000D 00    0.837016822000D 00   -0.813635703700D 00    0.132817656300D 00
 0.250000000000D 00    0.796355167200D 00   -0.812836073900D 00    0.165885062400D 00
 0.300000000000D 00    0.755731801300D 00   -0.812115943800D 00    0.198890221700D 00
 0.350000000000D 00    0.715140394400D 00   -0.811568762400D 00    0.231843469200D 00
 0.400000000000D 00    0.674570051000D 00   -0.811283765700D 00    0.264762524900D 00
 0.450000000000D 00    0.634005509500D 00   -0.811346426500D 00    0.297672165400D 00
 0.500000000000D 00    0.593427319600D 00   -0.811838883400D 00    0.809233561900D 00
```

T, KV, DELV

```
 0.500000000000D 00   -0.218411563800D 00    0.587533691D-02
 0.450000000000D 00   -0.177340916900D 00    0.455074176500D-02
 0.400000000000D 00   -0.136713714700D 00    0.426886610300D-02
 0.350000000000D 00   -0.964283680400D-01    0.433959154100D-02
 0.300000000000D 00   -0.563841424900D-01    0.466233959400D-02
 0.250000000000D 00   -0.164809066600D-01    0.513776502000D-02
 0.200000000000D 00    0.233811183100D-01    0.566725697100D-02
 0.150000000000D 00    0.633016082600D 00    0.615244286200D-02
 0.100000000000D 00    0.103380385200D 00    0.649469309900D-02
 0.500000000000D-01    0.143717667000D 00    0.659462479000D-02
-0.473579508900D-15    0.184414317700D 00    0.635160259100D-02
```

ITERATION= 1   TRIAL= 0   J= 0.809233561900D 00
ALPHA= 0.000000000000D 00   DJ= -0.295332466500D-04 |DHV|= 0.180882998800D-01

T,YV,J

0.00000000000D 00	0.10000000000D 01	-0.80921712140D 00	0.00000000000D 00
0.50000000000D-01	0.95953929120D 00	-0.80920939230D 00	0.32741379190D-01
0.10000000000D 00	0.91907933910D 00	-0.80918692050D 00	0.65481533620D-01
0.15000000000D 00	0.87862086600D 00	-0.80915038840D 00	0.98219294480D-01
0.20000000000D 00	0.83816454520D 00	-0.80910103880D 00	0.13095357240D 00
0.25000000000D 00	0.79771097950D 00	-0.80904040370D 00	0.16368339200D 00
0.30000000000D 00	0.75726069380D 00	-0.80897004860D 00	0.19640790430D 00
0.35000000000D 00	0.71681413910D 00	-0.80889132920D 00	0.22912638010D 00
0.40000000000D 00	0.67637170890D 00	-0.80880516000D 00	0.26183818330D 00
0.45000000000D 00	0.63593376700D 00	-0.80871179210D 00	0.29454272610D 00
0.50000000000D 00	0.59550068610D 00	-0.80861059890D 00	0.80921940830D 00

ITERATION= 1   TRIAL= 1   J= 0.80921940830D 00
ALPHA= 0.10026699160D 01   DJ= 0.12988927550D-05 |DHV|= 0.18088299880D-01

T,YV,J

0.00000000000D 00	0.10000000000D 01	-0.80948547620D 00	0.00000000000D 00
0.50000000000D-01	0.95952590020D 00	-0.80947462370D 00	0.32763055140D-01
0.10000000000D 00	0.91905307880D 00	-0.80943506050D 00	0.65524040570D-01
0.15000000000D 00	0.87858283460D 00	-0.80937229720D 00	0.98280853930D-01
0.20000000000D 00	0.83811618310D 00	-0.80929211790D 00	0.13103185160D 00
0.25000000000D 00	0.79765384910D 00	-0.80920034350D 00	0.16377586110D 00
0.30000000000D 00	0.75719627020D 00	-0.80910260850D 00	0.19651217490D 00
0.35000000000D 00	0.71674361180D 00	-0.80900414930D 00	0.22924052640D 00

(*continued*)

Table 5-5c Continued

```
 0.400000000D 00   0.6762957917D 00  -0.8089096021D 00   0.2619610495D 00
 0.450000000D 00   0.6358525152D 00  -0.8088228088D 00   0.2946742217D 00
 0.500000000D 00   0.5954133198D 00  -0.8087466305D 00   0.8092193808D 00

T,KV,DELV
 0.500000000D 00  -0.2133333107D 00  -0.5039732864D-03
 0.450000000D 00  -0.1729702936D 00  -0.4050533736D-03
 0.400000000D 00  -0.1326138104D 00  -0.3006109468D-03
 0.350000000D 00  -0.9226053757D-01  -0.1937091763D-03
 0.300000000D 00  -0.5190633834D-01  -0.8821837915D-04
 0.250000000D 00  -0.1154649443D-01  -0.1140597699D-04
 0.200000000D 00   0.2882406520D-01   0.1003409026D-03
 0.150000000D 00   0.6921053738D-01   0.1736108673D-03
 0.100000000D 00   0.1096180183D 00   0.2263018375D-03
 0.500000000D-01   0.1500512764D 00   0.2537760700D-03
-0.4735795089D-15  0.1905145238D 00   0.2518893678D-03

ITERATION= 2   TRIAL= 0    J= 0.8092193808D 00
ALPHA= 0.0000000000D 00   DJ= -0.6058374113D-07   |DHV|= 0.1007946573D-02

ITERATION LIMIT
```

5.3 Optimal Control Functions with a Variable Time Interval

The computer program which is described in this section applies to a class of forcing-function optimization problems that also includes parameter optimization and terminal equality constraints. Specifically, minimize

$$J = E(c) + \int_0^{t_f} F(x,u)dt + G(x_f) \qquad (5.15)$$

with respect to control vector $u(t)$ and parameter vector c where state vector $x(t)$ is defined by

$$\dot{x} = f(x, u), x(0) = \gamma + \Gamma c, g(x_f) = 0. \qquad (5.16)$$

Final time t_f is fixed. However, suitable selection of state vector $x(t)$ and parameter vector c results in the solution of an important class of variable end-time problems [1].

Description of the Basic Method

An extension of the Newton-Raphson method of successive substitutions is used to solve the optimization problem posed by equations (5.15) and (5.16). In particular, the program described subsequently in this section is a direct implementation of Theorem 6-7 in *Automated Design of Control Systems* [1]. The following sets of forward-time and backward-time equations are used in this implementation:

Forward-time equations

$$\left. \begin{array}{l} \dot{x}_0 = F(x, u), x_0(0) = E(c), J = x_0(t_f) + G(x_f) \\ \dot{x} = f(x, u), x(0) = \gamma + \Gamma c \\ c = \alpha d + e, \lambda = \alpha \mu + v, u = \alpha \delta + k - L\lambda - Kx \end{array} \right\} \qquad (5.17)$$

Backward-time equations

$$-\dot{p} = K'H_u + (F_x - K'F_u) + A'p, \ p(t_f) = L_{x_f}$$

$$-\dot{q} = -K'H_u + A'q; \ q(t_f) = 0$$

$$-\dot{\Omega} = H_{xx} + \Omega H_{px} + H'_{px}\Omega - K'H_{uu}K, \ \Omega(t_f) = L_{x_f x_f}$$

$$-\dot{U} = A'U, \ U(t_f) = L_{x_f \lambda}$$

$$-\dot{W} = L'H_{uu}L, \ W(t_f) = 0$$

$$-\dot{w} = L'H_{uu}\delta, \ w(t_f) = L_\lambda$$

$$H = F + p'f, \ L = G + \lambda'g, \ A = H_{px} - H_{pu}K$$

$$\delta = -H_{uu}^{-1}(H_u + H_{up}q), \ L = H_{uu}^{-1}H_{up}U, \ K = H_{uu}^{-1}(H_{ux} + H_{up}\Omega)$$

$$k = u + L\lambda + Kx, \ e = c, \ v = \lambda$$

$$d = -\{E_{cc} + \Gamma'[\Omega(0) + U(0)W^{-1}(0)U'(0)]\Gamma\}^{-1}\{E_c$$
$$+ \Gamma'[p(0) + q(0) + U(0)W^{-1}(0)w(0)]\}$$

$$\mu = W^{-1}(0)[w(0) + U'(0)\Gamma d] \qquad (5.18)$$

In order to prevent finite escape times in the solution to matrix Riccati equations appearing in equation (5.18) on suboptimal trajectories, a parameter ρ is introduced into matrices

$$H_{xx} = F_{xx} + \rho(p'f)_{xx}, \ H_{ux} = F_{ux} + \rho(p'f)_{ux}, \ H_{uu}$$
$$= F_{uu} + \rho(p'f)_{uu}. \qquad (5.19)$$

This parameter is selected via program input data on the interval $[0,1]$. Case $\rho = 1$ corresponds to the Newton-Raphson method of successive substitutions, whereas case $\rho = 0$ corresponds to an asymptotic stability constraint [10] which is employed on suboptimal trajectories in order to avoid computational difficulties in the backward-time solution of matrix Riccati variables Ω.

MIN5 and MIN5S

The Newton-Raphson method of successive substitutions outlined previously in this section is implemented by subprogram MIN5 which is summarized as follows:

MIN5 function: Newton-Raphson method of successive substitution for minimizing

$$J = E(c) + \int_0^{t_f} F(x, u)dt + G(x_f)$$

with respect to $u(t)$ and c, where

$$\dot{x} = f(x, u), x(0) = \gamma + \Gamma c, g(x_f) = 0$$

and

$$u = k - J\begin{bmatrix} x \\ \lambda \end{bmatrix}$$

MIN5 dummy argument list: None.

MIN5 input data list: C1/C2/C3/NI,NII,NP,MTAPE/NX,NU,NG,NC/DT,RHO/ MINV/MAXV/GAMM/ZV/KV/JM, where
  NI = maximum number of iterations.
  NII = number of integration increments.
  NP = number of integration increments omitted between output.
  MTAPE = 1 for constant z, k, J input on cards or 2 for time varying z, k, J input on I/O unit 3.
  $1 \leqslant NX \leqslant 7, 1 \leqslant NU \leqslant 3, 0 \leqslant NG \leqslant 3, 0 \leqslant NC \leqslant 1$.
  DT = length of integration increments.
  RHO = asymptotic stability parameter (between 0 and 1).
  MINV(NU+NG+NC) = vector of minimum magnitudes of gradients, with respect to u, λ, and c, requiring additional interations.
  MAXV(NU+NG+NC) = vector of maximum permissible magnitudes of increments in the elements of u, λ, and c.
  GAMM(NX,NC) = Γ.
  ZV(NX+NG+NC) = vector formed with x_o, λ, and c (either on cards or I/O unit 3).
  KV(NU) = k (either on cards or I/O unit 3).
  JM(NU,NX+NG) = J (either on cards or I/O unit 3).

MIN5 output data list: ZV/KV/JM, where
  ZV(NX+NG+NC) = vector formed with x_o, λ, and c (on I/O unit 4).
  KV(NU) = k (on I/O unit 4).
  JM(NU,NX+NG) = J(on I/O unit 4).

Subprograms called by MIN5: MI and MIN5S for the computation of
  E = E(c), DEV(NC) = E_c,
  DDEM(NC,NC) = E_{cc}, F = $F(y,t)$, FV(NX) = $f(y,t)$,
  G = $G(x)$, DGV(NX) = G_x, DDGM(NX,NX) = G_{xx},
  GV(NG) = $g(x)$, DGM(NG,NX) = g_x, DDGA(NG,NX,NX)
  = $(g_1)_{xx}/(g_2)_{xx}/\ldots$, DFV(NX+NU) = F_y,
  DDFM(NX+NU,NX+NU) = F_{yy}, DFM(NX,NX+NU) = f_y, and DDFA
  (NX,NX+NU,NX+NU) = $(f_1)_{yy}/(f_2)_{yy}/\ldots$, where

$$y = \begin{bmatrix} x \\ u \end{bmatrix}.$$

Subprogram MIN5 is listed in Table 5-6. User problems are specified for subprogram MIN5 by writing an appropriate subprogram MIN5S. Specifications for writing this subprogram as given as follows:

MIN5S function: Compute $E(c)$, E_c, E_{cc}, $F(y,t)$, $f(y,t)$, $G(x)$, G_x, G_{xx}, $g(x)$, g_x, $(g_1)_{xx}/(g_2)_{xx}/\ldots$, F_y, F_{yy}, f_y, and $(f_1)_{yy}/(f_2)_{yy}/\ldots$ for use with MIN5, where

$$y = \begin{bmatrix} x \\ u \end{bmatrix}.$$

MIN5S dummy argument list: MC,MG,MX,MY,CV,YV,T,E,DEV,DDEM,F,FV, G,DGV,DDGM,GV,DGM,DDGA,DFV,DDFM,DFM,DDFA,MODE, where
  CV(MC) = c.
  YV(MY) = y.
  T = t.
  E = $E(c)$.
  DEV(MC) = E_c.
  DDEM(MC,MC) = E_{cc}.
  F = $F(y,t)$.
  FV(MX) = $f(y,t)$.
  G = $G(x)$.
  DGV(MX) = G_x.
  DDGM(MX,MX) = G_{xx}.
  GV(MG) = $g(x)$.
  DGM(MG,MX) = g_x.

Table 5-6
Listing of MIN5

```
       SUBROUTINE MIN5
       REAL*8 SLA(NU,NL,NTAB),DDGA(NG,NX,NX),DDFA(NX,NY,NY),
C     1SKM(NU,NTAB),SYM(NY,NTAB),SDELM(NU,NTAB),DFM(NX,NY),DGM(NG,NX),
C     2DDEM(NC,NC),DDFM(NY,NY),DDGM(NX,NX),GAMM(NX,NC),AM(NX,NX),
C     3LM(NU,NL),TLM(NU,NL),HM(MH,MH),PM(NL,NR),DPM(NL,NR),
C     4SDPM(NL,NR),CV(NC),YV(NY),ZV(NZ),FV(NX),GV(NG),DEV(NC),DFV(NY)
C     5DGV(NX),KV(NU),DELV(NU),PV(NX),SFV(NX),DZV(NZ),SZV(NZ),
C     6MINV(ND),MAXV(ND),HV(NU),TDELV(NU),XV(NL),DPV(NX)
       REAL*8 SLA(3,10,401),DDGA(4,6,6),DDFA(6,9,9),SKM(3,401),
      1SYM(9,401),SDELM(3,401),DFM(6,9),DGM(4,6),DDEM(1,1),
      2DDFM(9,9),DDGM(6,6),GAMM(6,1),AM(6,6),LM(3,10),TLM(3,10),
      3HM(4,4),PM(10,11),DPM(10,11),SDPM(10,11),CV(1),YV(9),
      4ZV(11),FV(6),GV(4),DEV(1),DFV(9),DGV(6),KV(3),DELV(3),PV(6),
      5SFV(6),DZV(11),SZV(11),MINV(8),MAXV(8),HV(3),TDELV(3),
      6XV(10),DPV(6)
       REAL*8 B,C,D,E,F,G,G1,G2,G3,G4,G5,G6,G7,G8,DT,DTH,T,RHO,
      1ALP,DABS,DSQRT
       DATA MH/4/,MC/1/,MG/4/,MX/6/,MY/9/
       DATA E/0D0/,DEV/0D0/,DDEM/0D0/,F/0D0/,FV/6*0D0/,G/0D0/,
      1DGV/6*0D0/,DDGM/36*0D0/,GV/4*0D0/,DGM/24*0D0/,DDGA/144*0D0/,
      2DFV/9*0D0/,DDFM/81*0D0/,DFM/54*0D0/,DDFA/486*0D0/
       LOGICAL*1 COM(65)
 1000 FORMAT(65A1)
 1001 FORMAT(4I10)
```

(continued)

Table 5-6 Continued

```
1002 FORMAT(4D18.10)
1003 FORMAT(4X,65A1)
1004 FORMAT(4X,4D18.10)
1005 FORMAT('1')
1006 FORMAT('  NI,NII,NP,MTAPE')
1007 FORMAT('  NX,NU,NG,NC')
1008 FORMAT('  DT,RHO')
1009 FORMAT('  MAXV')
1010 FORMAT('  KV')
1011 FORMAT('  KM')
1012 FORMAT(///,'  ZV')
1013 FORMAT(///,'  T,YV,J')
1014 FORMAT(///,'  T,JM,KV,DELV')
1015 FORMAT('  GAMM')
1016 FORMAT(///,'  GV')
1017 FORMAT(///,'  ITERATION=',I2,'   ALPHA=',D18.10,
     1'  L=',D18.10)
1018 FORMAT(///,'  DESIGN COMPLETE')
1019 FORMAT(///,'  ITERATION LIMIT')
1020 FORMAT('  MINV')
     WRITE (6,1005)
     DO 1021 I=1,3
     READ (5,1000) COM
1021 WRITE (6,1003) COM
     READ (5,1001) NI,NII,NP,MTAPE
     READ (5,1001) NX,NU,NG,NC
```

```
51          NL=NX+NG
52          NR=NL+1
53          NE=NG+NC
54          NZ=NL+NC
55          NY=NX+NU
56          NK=NU+NG
57          ND=NK+NC
58          NTAB=NII+1
59          NIC=0
60          READ  (5,1002)  DT,RHO
61          READ  (5,1002)  (MINV(I),I=1,ND)
62          READ  (5,1002)  (MAXV(I),I=1,ND)
63          WRITE (6,1006)
64          WRITE (6,1001)  NI,NII,NP,MTAPE
65          WRITE (6,1007)
66          WRITE (6,1001)  NX,NU,NG,NC
67          WRITE (6,1008)
68          WRITE (6,1004)  DT,RHO
69          WRITE (6,1020)
70          WRITE (6,1004)  (MINV(I),I=1,ND)
71          WRITE (6,1009)
72          WRITE (6,1004)  (MAXV(I),I=1,ND)
73          IF (NC) 1023,1023,1022
74    1022  READ  (5,1002)  ((GAMM(I,J),I=1,NX),J=1,NC)
75          WRITE (6,1015)
76          WRITE (6,1004)  ((GAMM(I,J),I=1,NX),J=1,NC)
77    1023  GO TO (1024,1026),MTAPE
```

Table 5-6 Continued

```
78   1024 READ  (5, 1002) (SZV(I),I=1,NZ)
79        READ  (5, 1002) (KV(I),I=1,NU)
80        READ  (5, 1002) ((LM(I,J),I=1,NU),J=1,NX)
81        WRITE (6, 1010)
82        WRITE (6, 1004) (KV(I),I=1,NU)
83        WRITE (6, 1011)
84        WRITE (6, 1004) ((LM(I,J),I=1,NU),J=1,NX)
85        DO 1025 K=1,NTAB
86        DO 1025 I=1,NU
87        SKM(I,K)=KV(I)
88        DO 1025 J=1,NX
89   1025 SLA(I,J,K)=LM(I,J)
90        GO TO 1027
91   1026 READ (3) (SZV(I),I=1,NZ),((SKM(I,K),I=1,NU),((SLA(I,J,K),
     1I=1,NU),J=1,NX),K=1,NTAB)
93        REWIND 3
94   1027 DO 1028 I=1,NZ
95   1028 DZV(I)=0D0
96        DO 1031 K=1,NTAB
97        DO 1031 I=1,NU
98        IF (NG) 1031,1031,1029
99   1029 DO 1030 J=1,NG
100  1030 SLA(I,J+NX,K)=0D0
101  1031 SDELM(I,K)=0D0
102       ALP=0D0
103       D=DSQRT(5D-1)
```

```
104          G1=DT*(1D0-D)
105          G2=DT*(1D0+D)
106          D=3D0*D
107          G6=D-2D0
108          G8=-(D+2D0)
109          G3=DT/6D0
110          G4=-DT/3D0
111          D=DSQRT(2D0)
112          G5=2D0-D
113          G7=2D0+D
114          DTH=5D-1*DT
115          CALL MIN5S(MC,MG,MX,MY,CV,YV,T,E,DEV,DDEM,F,FV,G,DGV,DDGM,
116         1GV,DGM,DDGA,DFV,DDFM,DFM,DDFA,1)
117     1032 NIP=NP
118          NIIC=0
119          I=0
120          MGILL=1
121          T=0D0
122          DO 1033 I=1,NZ
123     1033 ZV(I)=ALP*DZV(I)+SZV(I)
124          IF (NC) 1036,1036,1034
125     1034 DO 1035 I=1,NC
126     1035 CV(I)=ZV(I+NL)
127     1036 CALL MIN5S(MC,MG,MX,MY,CV,YV,T,E,DEV,DDEM,F,FV,G,DGV,DDGM,
128         1GV,DGM,DEGA,DFV,DDFM,DFM,DDFA,2)
129          DO 1037 I=1,NL
130     1037 XV(I)=ZV(I)
```

(continued)

Table 5-6 Continued

```
131          C=E
132          DO 1038 I=1,NU
133          DELV(I)=SDELM(I,1)
134          KV(I)=SKM(I,1)
135          DO 1038 J=1,NL
136  1038    LM(I,J)=SLA(I,J,1)
137          WRITE (6,1012)
138          WRITE (6,1004) (ZV(I),I=1,NZ)
139          WRITE (6,1013)
140          DO 1040 I=1,NX
141  1039    YV(I)=XV(I)
142  1040    DO 1042 I=1,NU
143          D=KV(I)+ALP*DELV(I)
144          DO 1041 J=1,NL
145  1041    D=D-LM(I,J)*XV(J)
146  1042    YV(I+NX)=D
147          CALL MIN5S(MC,MG,MX,MY,CV,YV,T,E,DEV,DDEM,F,FV,G,DGV,DDGM,
148         1GV,DGM,DDGA,DFV,DDFM,DFM,DDFA,3)
149          GO TO (1043,1049,1051,1054),MGILL
150  1043    NIIC=NIIC+1
151          DO 1044 I=1,NY
152  1044    SYM(I,NIIC)=YV(I)
153          NIP=NIP+1
154          IF (NIP .LE. NP) GO TO 1046
155          IF (L) 1045,1045,1056
156  1045    WRITE (6,1004) T,(YV(I),I=1,NY),C
```

```
157            NIP=0
158       1046 DO 1047 I=1,NX
159            SFV(I)=FV(I)
160       1047 XV(I)=XV(I)+DTH*FV(I)
161            B=F
162            C=C+DTH*F
163            T=T+DTH
164            M=NIIC+1
165            DO 1048 I=1,NU
166            DELV(I)=5D-1*(SDELM(I,NIIC)+SDELM(I,M))
167            KV(I)=5D-1*(SKM(I,NIIC)+SKM(I,M))
168            DO 1048 J=1,NL
169       1048 LM(I,J)=5D-1*(SLA(I,J,NIIC)+SLA(I,J,M))
170            MGILL=2
171            GO TO 1039
172       1049 DO 1050 I=1,NX
173            XV(I)=XV(I)+G1*(FV(I)-SFV(I))
174       1050 SFV(I)=G5*FV(I)+G6*SFV(I)
175            C=C+G1*(F-B)
176            B=G5*F+G6*B
177            MGILL=3
178            GO TO 1039
179       1051 DO 1052 I=1,NX
180            XV(I)=XV(I)+G2*(FV(I)-SFV(I))
181       1052 SFV(I)=G7*FV(I)+G8*SFV(I)
182            C=C+G2*(F-B)
183            B=G7*F+G8*B
```

(continued)

Table 5-6 Continued

```
184            T=T+DTH
185            DO 1053 I=1,NU
186            DELV(I)=SDELM(I,M)
187            KV(I)=SKM(I,M)
188            DO 1053 J=1,NL
189      1053  LM(I,J)=SLA(I,J,M)
190            MGILL=4
191            GO TO 1039
192      1054  DO 1055 I=1,NX
193      1055  XV(I)=XV(I)+G3*FV(I)+G4*SFV(I)
194            C=C+G3*F+G4*B
195            MGILL=1
196            IF (NIIC .LT. NII) GO TO 1039
197            L=1
198            NIP=NP
199            GO TO 1039
200      1056  CALL MIN5S(MC,MG,MX,MY,CV,YV,T,E,DEV,DDEM,F,FV,G,DGV,DDGM,
201           1GV,DGM,DDGA,DFV,DDFM,DFM,DDFA,4)
202            C=C+G
203            WRITE (6,1004)  T,(YV(I),I=1,NY),C
204            IF (NG) 1059,1059,1057
205      1057  WRITE (6,1016)
206            WRITE (6,1004)  (GV(I),I=1,NG)
207            DO 1058 I=1,NG
208      1058  C=C+ZV(I+NX)*GV(I)
209      1059  WRITE (6,1017) NIC,ALP,C
```

OPTIMIZATION ON A FINITE TIME INTERVAL

(continued)

```
210         IF (NIC .GE. NI)  GO TO 1161
211         WRITE (6,1014)
212         NIC=NIC+1
213         NIP=NP
214         L=0
215         MSTOP=0
216         DO 1065 I=1,NX
217         PM(I,NR)=0D0
218         D=DGV(I)
219         IF (NG)  1062,1062,1060
220    1060 DO 1061 J=1,NG
221    1061 D=D+ZV(J+NX)*DGM(J,I)
222    1062 PV(I)=D
223         DO 1065 J=I,NX
224         D=0D0
225         IF (NG)  1065,1065,1063
226    1063 DO 1064 K=1,NG
227    1064 D=D+ZV(K+NX)*DDGA(K,I,J)
228    1065 PM(I,J)=DDGM(I,J)+RHO*D
229         IF (NG)  1071,1071,1066
230    1066 DO 1070 J=1,NG
231         D=GV(J)
232         IF (DABS(D) .GT. MINV(J+NU))  GO TO 1067
233         D=0D0
234         GO TO 1068
235    1067 MSTOP=1
236    1068 PM(J+NX,NR)=D
```

Table 5-6 Continued

```
237            DO 1069 I=1,NX
238  1069   PM(I,J+NX)=DGM(J,I)
239            DO 1070 I=1,J
240  1070   PM(I+NX,J+NX)=0D0
241  1071   CALL MIN5S(MC,MG,MX,MY,CV,YV,T,E,DEV,DDEM,F,FV,G,DGV,DDGM,
242           1GV,DGM,DEGA,DFV,DDFM,DFM,DDFA,5)
243  1072   DO 1073 I=1,NX
244            DO 1073 J=I,NX
245  1073   PM(J,I)=PM(I,J)
246            DO 1082 I=1,NU
247            D=DFV(I+NX)
248            DO 1074 J=1,NX
249  1074   D=D+DFM(J,I+NX)*PV(J)
250            IF (DABS(D) .GT. MINV(I)) GO TO 1075
251            D=0D0
252            GO TO 1076
253  1075   MSTOP=1
254  1076   HV(I)=D
255            DO 1078 J=1,NX
256            D=D+DFM(J,I+NX)*PM(J,NR)
257            C=0D0
258            B=DDFM(J,I+NX)
259            DO 1077 K=1,NX
260            C=C+PV(K)*DDFA(K,J,I+NX)
261  1077   B=B+DFM(K,I+NX)*PM(K,J)
262  1078   TLM(I,J)=B+RHO*C
```

```
263             IF (NG) 1082,1082,1079
264        1079 DO 1081 J=1,NG
265             C=0D0
266             DO 1080 K=1,NX
267        1080 C=C+DFM(K,I+NX)*PM(K,J+NX)
268        1081 TLM(I,J+NX)=C
269        1082 TDELV(I)=D
270             DO 1084 I=1,NU
271             DO 1084 J=1,NU
272             D=0D0
273             DO 1083 K=1,NX
274        1083 D=D+PV(K)*DDFA(K,I+NX,J+NX)
275             HM(I,J)=DDFM(I+NX,J+NX)+RHO*D
276        1084 HM(J,I)=HM(I,J)
277             CALL MI(MH,NU,HM)
278             DO 1087 I=1,NU
279             D=0D0
280             DO 1085 J=1,NU
281        1085 D=D-HM(I,J)*TDELV(J)
282             EELV(I)=D
283             DO 1087 J=1,NL
284             D=0D0
285             DO 1086 K=1,NU
286        1086 D=D+HM(I,K)*TLM(K,J)
287        1087 LM(I,J)=D
288             DO 1090 I=1,NX
289             D=0D0
290             C=DFV(I)
```

(continued)

Table 5-6 Continued

```
291              DO 1088  J=1,NU
292              D=D+LM(J,I)*HV(J)
293        1088  C=C-LM(J,I)*DFV(J+NX)
294              DPV(I)=D+C
295              DPM(I,NR)=-D
296              DO 1090  J=1,NX
297              D=DFM(I,J)
298              DO 1089  K=1,NU
299        1089  D=D-DFM(I,K+NX)*LM(K,J)
300        1090  AM(I,J)=D
301              DO 1094  I=1,NX
302              D=DPV(I)
303              C=DPM(I,NR)
304              DO 1091  J=1,NX
305              D=D+AM(J,I)*PV(J)
306        1091  C=C+AM(J,I)*PM(J,NR)
307              DPV(I)=D
308              DPM(I,NR)=C
309              DO 1094  J=I,NX
310              D=0D0
311              C=DDFM(I,J)
312              DO 1092  K=1,NX
313              D=D+PV(K)*DDFA(K,I,J)
314        1092  C=C+PM(I,K)*DFM(K,J)+DFM(K,I)*PM(K,J)
315              DO 1093  K=1,NU
316        1093  C=C-LM(K,I)*TLM(K,J)
```

```
317  1094  DPM(I,J)=C+RHO*D
318        IF (NG) 1101,1101,1095
319  1095  DO 1100 I=1,NG
320        D=0D0
321        DO 1096 J=1,NU
322  1096  D=D-LM(J,I+NX)*TDELV(J)
323        DPM(I+NX,NR)=D
324        DO 1098 J=1,NX
325        D=0D0
326        DO 1097 K=1,NX
327  1097  D=D+AM(K,J)*PM(K,I+NX)
328  1098  DPM(J,I+NX)=D
329        DO 1100 J=I,NG
330        D=0D0
331        DO 1099 K=1,NU
332  1099  D=D+LM(K,I+NX)*TIM(K,J+NX)
333  1100  DPM(I+NX,J+NX)=D
334  1101  GO TO (1102,1110,1113,1118),MGILL
335  1102  DO 1104 I=1,NU
336        D=YV(I+NX)
337        SDELM(I,NIIC)=DELV(I)
338        DO 1103 J=1,NL
339        D=D+LM(I,J)*XV(J)
340  1103  SLA(I,J,NIIC)=LM(I,J)
341        KV(I)=D
342  1104  SKM(I,NIIC)=D
343        NIP=NIP+1
344        IF (NIP .LE. NP) GO TO 1106
```

(continued)

Table 5-6 Continued

```
345         WRITE (6,1004) T,((LM(I,J),I=1,NU),J=1,NL),(KV(I),I=1,NU),
346        1(DELV(I),I=1,NU)
347         IF (L) 1105,1105,1121
348   1105 NIP=0
349   1106 DO 1107 I=1,NX
350        SFV(I)=DPV(I)
351   1107 PV(I)=PV(I)+DTH*DPV(I)
352        DO 1108 I=1,NL
353        DO 1108 J=I,NR
354        SDPM(I,J)=DPM(I,J)
355   1108 PM(I,J)=PM(I,J)+DTH*DPM(I,J)
356        T=T-DTH
357        M=NIIC-1
358        DO 1109 I=1,NY
359   1109 YV(I)=5D-1*(SYM(I,M)+SYM(I,NIIC))
360        NIIC=M
361        MGILL=2
362        GO TO 1071
363   1110 DO 1111 I=1,NX
364        PV(I)=PV(I)+G1*(DPV(I)-SFV(I))
365   1111 SFV(I)=G5*DPV(I)+G6*SFV(I)
366        DO 1112 I=1,NL
367        DO 1112 J=I,NR
368        PM(I,J)=PM(I,J)+G1*(DPM(I,J)-SDPM(I,J))
369   1112 SDPM(I,J)=G5*DPM(I,J)+G6*SDPM(I,J)
370        MGILL=3
```

```
            GO TO 1072
371    1113 DO 1114 I=1,NX
372         PV(I)=PV(I)+G2*(DPV(I)-SFV(I))
373    1114 SFV(I)=G7*DPV(I)+G8*SFV(I)
374         DO 1115 I=1,NL
375         DO 1115 J=I,NR
376         PM(I,J)=PM(I,J)+G2*(DPM(I,J)-SDPM(I,J))
377    1115 SDPM(I,J)=G7*DPM(I,J)+G8*SDPM(I,J)
378         T=T-DTH
379         DO 1116 I=1,NY
380    1116 YV(I)=SYM(I,NIIC)
381         DO 1117 I=1,NX
382    1117 XV(I)=YV(I)
383         MGILL=4
384         GO TO 1071
385    1118 DO 1119 I=1,NX
386    1119 PV(I)=PV(I)+G3*DPV(I)+G4*SFV(I)
387         DO 1120 I=1,NL
388         DO 1120 J=I,NR
389    1120 PM(I,J)=PM(I,J)+G3*DPM(I,J)+G4*SDPM(I,J)
390         MGILL=1
391         IF (NIIC .GT. 1) GO TO 1072
392         L=1
393         NIP=NP
394         GO TO 1072
395    1121 IF (NG) 1131,1131,1122
396    1122 DO 1123 I=1,NG
397         DO 1123 J=I,NG
```

(continued)

Table 5-6 Continued

```
399            HM(I,J)=PM(I+NX,J+NX)
400       1123 HM(J,I)=HM(I,J)
401            CALL MI(MH,NG,HM)
402            DO 1125 I=1,NG
403            D=0D0
404            DO 1124 J=1,NG
405       1124 D=D+HM(I,J)*PM(J+NX,NR)
406       1125 DZV(I+NX)=D
407            IF (NC) 1154,1154,1126
408       1126 DO 1128 I=1,NG
409            DO 1128 J=1,NX
410            D=0D0
411            DO 1127 K=1,NG
412       1127 D=D+HM(I,K)*PM(J,K+NX)
413       1128 DPM(I,J)=D
414            DO 1130 I=1,NG
415            DO 1130 J=1,NC
416            D=0D0
417            DO 1129 K=1,NX
418       1129 D=D+DPM(I,K)*GAMM(K,J)
419       1130 SDPM(I,J)=D
420       1131 IF (NC) 1154,1154,1132
421       1132 DO 1137 I=1,NX
422            D=PM(I,NR)
423            IF (NG) 1135,1135,1133
424       1133 DO 1134 J=1,NG
```

```
425  1134  D=D+PM(I,J+NX)*DZV(J+NX)
426  1135  PM(I,NR)=D
427        DO 1137 J=1,NC
428        D=0D0
429        DO 1136 K=1,NX
430  1136  D=D+PM(I,K)*GAMM(K,J)
431  1137  DPM(I,J)=D
432        DO 1143 I=1,NC
433        D=DEV(I)
434        B=0D0
435        DO 1141 J=1,NX
436        C=DPM(J,I)
437        IF (NG) 1140,1140,1138
438  1138  DO 1139 K=1,NG
439  1139  C=C+PM(J,K+NX)*SDPM(K,I)
440  1140  DPM(J,I)=C
441        B=B+GAMM(J,I)*PM(J,NR)
442  1141  D=D+GAMM(J,I)*PV(J)
443        IF (DABS(D) .GT. MINV(I+NK)) GO TO 1142
444        D=0D0
445        GO TO 1143
446  1142  MSTOP=1
447  1143  SZV(I)=D+B
448        DO 1145 I=1,NC
449        DO 1145 J=I,NC
450        D=DDEM(I,J)
451        DO 1144 K=1,NX
452  1144  D=D+GAMM(K,I)*DPM(K,J)
453        HM(I,J)=D
```

(continued)

Table 5-6 Continued

```
454  1145  HM(J,I)=D
455        CALL MI(MH,NC,HM)
456        DO 1147 I=1,NC
457        D=0D0
458        DO 1146 J=1,NC
459  1146  D=D-HM(I,J)*SZV(J)
460  1147  DZV(I+NL)=D
461        IF (NG) 1151,1151,1148
462  1148  DO 1150 I=1,NG
463        D=DZV(I+NX)
464        DO 1149 J=1,NC
465  1149  D=D+SDPM(I,J)*DZV(J+NL)
466  1150  DZV(I+NX)=D
467  1151  DO 1153 I=1,NX
468        D=0D0
469        DO 1152 J=1,NC
470  1152  D=D+GAMM(I,J)*DZV(J+NL)
471  1153  DZV(I)=D
472  1154  IF (MSTOP) 1155,1155,1156
473  1155  WRITE (6,1018)
474        ALP=0D0
475        GO TO 1162
476  1156  ALP=1D0
477        DO 1157 I=1,NZ
478  1157  SZV(I)=ZV(I)
479        DO 1158 I=1,NU
```

```
480              D=MAXV(I)
481              DO 1158 J=1,NTAB
482              C=DABS(SDELM(I,J))
483              IF (D .LT. ALP*C)   ALP=D/C
484         1158 CONTINUE
485              IF (NE)  1032,1032,1159
486         1159 DO 1160 I=1,NE
487              D=MAXV(I+NU)
488              C=DABS(DZV(I+NX))
489              IF (D .LT. ALP*C)   ALP=D/C
490         1160 CONTINUE
491              GO TO 1032
492         1161 WRITE (6,1019)
493         1162 DO 1165 K=1,NTAB
494              DO 1165 I=1,NU
495              D=SKM(I,K)+ALP*SDELM(I,K)
496              IF (NG)  1165,1165,1163
497         1163 DO 1164 J=1,NG
498         1164 D=D-SLA(I,J+NX,K)*ZV(J+NX)
499         1165 SKM(I,K)=D
500              WRITE (4) (ZV(I),I=1,NZ),((SKM(I,K),I=1,NU),((SLA(I,J,K),
501             1I=1,NU),J=1,NX),K=1,NTAB)
502              END FILE 4
503              REWIND 4
504              WRITE (6,1005)
505              RETURN
506              END
```

DDGA(MG,MX,MX) = $(g_1)_{xx}/(g_2)_{xx}/\ldots$.
DFV(MY) = F_y.
DDFM(MY,MY) = F_{yy}.
DFM(MX,MY) = f_y.
DDFA(MX,MY,MY) = $(f_1)_{yy}/(f_2)_{yy}/\ldots$.
MODE = 1 for program initialization, 2 for the computation of E, DEV, and DDEM, 3 for the computation of F and FV, 4 for the computation of G, DGV, DDGM, GV, DGM, and DDGA, and 5 for the computation of DFV, DDFM, DFM, and DDFA.

Example 8A

An example of the use of subprogram MIN5 is provided by the problem of computing optimal control function $u_1(t)$ and optimal final time t_f corresponding to the minimization of

$$J = t_f + \int_0^{t_f} \frac{1}{2} u_1^2 dt + \frac{1}{2} x_1^2(t_f)$$

where

$$\dot{x}_1 = u_1, x_1(0) = 1, x_1(t_f) = 0.$$

This optimization problem can be restated in the form given in Section 6.9 of *Automated Design of Control Systems* [1], and then subprogram MIN5S can be written in a straightforward manner using

$$E = c_1 t_f, F = \frac{1}{2} y_2 y_3^2, G = \frac{1}{2} y_1^2,$$

$$f_1 = y_2 y_3, f_2 = 0, g_1 = y_1.$$

and

$$\Gamma = \begin{bmatrix} 0 \\ 1 \end{bmatrix}.$$

Calling program MIN5C and subprogram MIN5S are listed in Table 5-7a. A listing of program input cards for this example is given in Table 5-7b. Finally, program printed output is listed in Table 5-7c for this example. Iterations given in Table 5-7 are performed with parameter $\rho = 1$ and illustrate quadratic convergence for this variable end-time problem.

Table 5-7a
Listing of User-supplied Programs for Example 8A

```
1   C    MIN5C  EXAMPLE 8A
2          CALL MIN5
3          STOP
4          END
5          SUBROUTINE MIN5S(MC,MG,MX,MY,CV,YV,T,E,DEV,
6         1DDEM,F,FV,G,DGV,DDGM,GV,DGM,DDGA,DFV,DDFM,
7         2DFM,DDFA,MODE)
8          REAL*8 DDGA(MG,MX,MX),DDFA(MX,MY,MY),DDEM(MC,MC),
9         1DDGM(MX,MX),DGM(MG,MX),DDFM(MY,MY),DFM(MX,MY),
10        2CV(MC),YV(MY),DEV(MC),FV(MX),DGV(MX),GV(MG),
11        3DFV(MY),T,E,F,G
12         REAL*8 TF
13   1000 FORMAT (D18.10)
14   1001 FORMAT (//,' TF=',D18.10)
15         GO TO (1002,1003,1004,1005,1006),MODE
16   1002 READ (5,1000) TF
17         WRITE (6,1001) TF
18         DDFA(1,2,3)=1.D0
19         DGM(1,1)=1.D0
20         DDGM(1,1)=1.D0
21         DEV(1)=TF
22         RETURN
23   1003 E=CV(1)*TF
24         RETURN
```

(continued)

Table 5-7a Continued

```
25    1004 FV(1)=YV(2)*YV(3)
26         F=FV(1)*YV(3)
27         RETURN
28    1005 GV(1)=YV(1)
29         DGV(1)=YV(1)
30         G=.5D0*YV(1)**2
31         RETURN
32    1006 DFV(2)=YV(3)**2
33         DDFM(2,3)=2.D0*YV(3)
34         DDFM(3,3)=2.D0*YV(2)
35         DFV(3)=DDFM(3,3)*YV(3)
36         DFM(1,2)=YV(3)
37         DFM(1,3)=YV(2)
38         RETURN
39         END
```

Table 5-7b
Listing of User-supplied Input Data for Example 8A

1	MIN5(MI,MIN5S)		EXAMPLE 8A				
2	MERRIAM						
3	UNIVERSITY OF ROCHESTER						
4		3	50		9		
5		2	1		1	1	
6	.02		1.				
7	0.		0.	0.			
8	5.		5.	5.			.9
9	0.		1.		2.		
10	1.		.9				
11	-1.						
12	0.		0.	0.			
13	1.						

Table 5-7c
Printed Output from MIN5 for Example 8A

```
MIN5(MI,MIN5S)   EXAMPLE 8A
MERRIAM
UNIVERSITY OF ROCHESTER
NI,NII,NP,MTAPE
  3      50       9      1
NX,NU,NG,NC
  2       1       1      1
DT,RHO
  0.2000000000D-01    0.1000000000D 01
MINV
  0.0000000000D 00    0.0000000000D 00    0.0000000000D 00
MAXV
  0.5000000000D 01    0.5000000000D 01    0.5000000000D 01
GAMM
  0.0000000000D 00    0.1000000000D 01
KV
 -0.1000000000D 01
KM
  0.0000000000D 00    0.0000000000D 00

TF=   0.1000000000D 01

ZV
  0.1000000000D 01    0.9000000000D 00    0.2000000000D 01    0.9000000000D 00
```

T,YV,J
0.0000000000D 00   0.1000000000D 01   0.9000000000D 00   -0.1000000000D 01
0.9000000000D 00   0.8200000000D 00   0.9000000000D 00   -0.1000000000D 01
0.2000000000D 01   0.6400000000D 00   0.9000000000D 00   -0.1000000000D 01
0.1080000000D 01   0.6400000000D 00   0.9000000000D 00   -0.1000000000D 01
0.4000000000D 00   0.4600000000D 00   0.9000000000D 00   -0.1000000000D 01
0.1260000000D 01   0.2800000000D 00   0.9000000000D 00   -0.1000000000D 01
0.6000000000D 00   0.2800000000D 00   0.9000000000D 00   -0.1000000000D 01
0.1440000000D 01   0.1000000000D 00   0.9000000000D 00   -0.1000000000D 01
0.8000000000D 00   0.1000000000D 00   0.9000000000D 00   -0.1000000000D 01
0.1620000000D 01
0.1000000000D 01
0.1805000000D 01

GV
0.1000000000D 00

ITERATION= 0   ALPHA= 0.0000000000D 00   L= 0.2005000000D 01

T,JM,KV,DELV
0.1000000000D 01   0.5000000000D 00   0.5555555556D-01   0.5000000000D 00
0.1000000000D 00   -0.5000000000D-01
0.8000000000D 00   0.4587155963D 00   -0.4077471964D-01   0.4587155963D 00
0.9174311997D-02   -0.4587155963D-01

(continued)

Table 5-7c Continued

0.6000000000D 00	0.4237288136D 00	−0.1224105461D 00	0.4237288136D 00
−0.6779661006D−01	−0.4237288136D−01		
0.4000000000D 00	0.3937007874D 00	−0.1924759404D 00	0.3937007874D 00
−0.1338582676D 00	−0.3937007874D−01		
0.2000000000D 00	0.3676470589D 00	−0.2532679738D 00	0.3676470589D 00
−0.1911764705D 00	−0.3676470589D−01		
−0.1301042607D−14	0.3448275862D 00	−0.3065134099D 00	0.3448275862D 00
−0.2413793102D 00	−0.3448275862D−01		

ZV

| 0.1000000000D 01 | 0.9954545455D 00 | 0.1999494950D 01 | 0.9954545455D 00 |

T,YV,J

0.0000000000D 00	0.1000000000D 01	0.9954545455D 00	−0.1005050505D 01
0.9954545455D 00			
0.2000000000D 00	0.7999054915D 00	0.9954545455D 00	−0.1005015759D 01
0.1196557712D 01			
0.4000000000D 00	0.5998179003D 00	0.9954545455D 00	−0.1004978812D 01
0.1397646975D 01			
0.6000000000D 00	0.3997376468D 00	0.9954545455D 00	−0.1004939338D 01
0.1598721490D 01			
0.8000000000D 00	0.1996652077D 00	0.9954545455D 00	−0.1004896931D 01
0.1799780298D 01			
0.1000000000D 01	−0.3988720551D−03	0.9954545455D 00	−0.1004851069D 01

0.20008223385D 01

GV
 -0.39887205551D-03

ITERATION= 1   ALPHA= 0.10000000000D 01   L= 0.20000024842D 01

T,JM,KV,DELV
  0.10000000000D 01   0.50000000000D 00 -0.53272455053D-02   0.50000000000D 00
 -0.10606060661D-01   0.53030303030D-02 -0.96278888382D-01   0.45473336090D 00
 -0.80000000000D 00   0.45473336090D 00 -0.17213335850D 00   0.41698256260D 00
 -0.10070669370D-02   0.48666871130D-02
 -0.60000000000D 00   0.41698256260D 00 -0.23636271380D 00   0.38501925100D 00
 -0.17585231540D 00   0.45032930510D-02
 -0.40000000000D 00   0.38501925100D 00 -0.29144921330D 00   0.35760728220D 00
 -0.23948166360D 00   0.41960421280D-02
 -0.20000000000D 00   0.35760728220D 00 -0.33921585950D 00   0.33383915030D 00
 -0.29405421990D 00   0.39329171670D-02
 -0.13010426070D-14   0.33383915030D 00
 -0.34137562920D 00   0.37050982130D-02

ZV
  0.10000000000D 01   0.10000063580D 01   0.19999706680D 01   0.10000063580D 01

(continued)

Table 5-7c Continued

```
T,YV,J
0.000000000D 00    0.1000000000D 01    0.1000006358D 01  -0.9999601735D 00
0.1000006358D 01   0.8000049379D 00    0.1000006358D 01  -0.9999619033D 00
0.2000000000D 00   0.6000092769D 00    0.1000006358D 01  -0.9999637360D 00
0.1199995210D 01
0.4000000000D 00   0.4000129347D 00    0.1000006358D 01  -0.9999656824D 00
0.1399985261D 01
0.6000000000D 00   0.2000158051D 00    0.1000006358D 01  -0.9999677535D 00
0.1599976674D 01
0.8000000000D 00   0.1774684283D-04    0.1000006358D 01  -0.9999699589D 00
0.1799969662D 01
0.1000000000D 01
0.1999964507D 01

GV
0.1774684283D-04

ITERATION= 2    ALPHA=  0.1000000000D 01    L=  0.2000000000D 01

T,JM,KV,DELV
0.1000000000D 01   0.5000000000D 00    0.2424846428D-04   0.5000000000D 00
0.4849723689D-04  -0.2424861844D-04
0.8000000000D 00   0.4545451919D 00   -0.9088205796D-01   0.4545451919D 00
```

244

OPTIMIZATION ON A FINITE TIME INTERVAL

```
-0.9085711575D-01   -0.2414822390D-04   -0.1666399997D 00    0.4166662252D 00
 0.6000000000D 00    0.4166662252D 00
-0.1666116331D 00   -0.2411981469D-04   -0.2307370841D 00    0.3846148204D 00
 0.4000000000D 00    0.3846148204D 00
-0.2307114674D 00   -0.2413523846D-04   -0.2856798077D 00    0.3571422085D 00
 0.2000000000D 00    0.3571422085D 00
-0.2856540556D 00   -0.2417788630D-04   -0.3332966623D 00    0.3333326270D 00
-0.1301042607D-14    0.3333326270D 00
-0.3332708510D 00   -0.2423761146D-04

ZV
 0.1000000000D 01    0.9999889230D 00    0.1999999999D 01    0.9999889230D 00

T,YV,J
 0.0000000000D 00    0.0000000000D 00    0.9999999990D 00   -0.9999999990D 00
 0.9999889230D 00    0.1000000000D 01
 0.2000000000D 00    0.8000006284D 00    0.9999999981D 00   -0.9999999981D 00
 0.1199989882D 01
 0.4000000000D 00    0.6000010129D 00    0.9999999967D 00   -0.9999999967D 00
 0.1399991328D 01
 0.6000000000D 00    0.4000010924D 00    0.9999999944D 00   -0.9999999944D 00
 0.1599993384D 01
 0.8000000000D 00    0.2000007840D 00    0.9999999908D 00   -0.9999999908D 00
 0.1799996217D 01
 0.1000000000D 01   -0.2846336613D-07    0.9999999849D 00   -0.9999999849D 00
 0.2000000057D 01
```

(*continued*)

Table 5-7c Continued

GV
−0.2846336613D−07

ITERATION= 3   ALPHA= 0.1000000000D 01   L= 0.2000000000D 01

ITERATION LIMIT

5.4 Linearized Optimal Control with a Variable Time Interval

Once an extremal has been obtained using the Newton-Raphson method of solving the corresponding two-point boundary problem, linearized optimal control can be implemented using the control equation appearing in the forward-time iteration equations. Simulation of this control system is often an important step in ascertaining the feasibility of implementing a feedback system based on optimization.

Specifically, linearized optimal control function $\mathbf{u}(t)$ and parameter $\mathbf{c}$ are sought for minimizing

$$J = E(\mathbf{c}) + \int_0^{t_f} F(\mathbf{x}, \mathbf{u}) dt + G(\mathbf{x}_f), \tag{5.20}$$

where state vector $\mathbf{x}(t)$ is defined by

$$\dot{\mathbf{x}} = \mathbf{f}(\mathbf{x}, \mathbf{u}), \mathbf{x}(0) = \mathbf{x}_n + \Delta\delta. \tag{5.21}$$

Vector $\mathbf{x}_n$ is the nominal initial value of the state vector that has been determined as the solution to the optimization problem defined by equations (5.15) and (5.16). Vector δ is a vector that causes perturbations about the nominal optimal trajectory.

Description of the Basic Method

Linearized optimal control for the problem posed by equations (5.20) and (5.21) can be derived by a direct extension of the method presented in Section 6.7 of *Automated Design of Control Systems* [1]. The following set of forward-time equations constitute a simulation of linearized optimal control; the following set of backward-time equations specify the linear optimal control system, and they are usually solved with vectors $\mathbf{x}_n(t)$ and $\mathbf{u}_n(t)$ taken from the nominal optimal trajectory.

Forward-time equations

$$\left. \begin{array}{l} \dot{x}_0 = F(\mathbf{x}, \mathbf{u}), x_0(0) = E(\mathbf{c}), J = x_0(t_f) + G(\mathbf{x}_f) \\ \dot{\mathbf{x}} = \mathbf{f}(\mathbf{x}, \mathbf{u}), \mathbf{x}(0) = \mathbf{x}_n + \mathbf{M}\delta \\ \mathbf{c} = \mathbf{c}_n + \mathbf{H}\delta, \boldsymbol{\ell} = \mathbf{D}\delta, \mathbf{u} = \mathbf{k} - \mathbf{L}\boldsymbol{\ell} - \mathbf{K}\mathbf{x} \end{array} \right\} \tag{5.22}$$

Backward-time equations

$$
\begin{aligned}
&-\dot{\mathbf{p}} = (F_x - \mathbf{K}'F_u) + \mathbf{A}'\mathbf{p}, \mathbf{p}(t_f) = L_{\mathbf{x}_f} \\
&-\dot{\boldsymbol{\Omega}} = H_{xx} + \boldsymbol{\Omega} H_{px} + H_{px}'\boldsymbol{\Omega} - \mathbf{K}'H_{uu}\mathbf{K}, \boldsymbol{\Omega}(t_f) = L_{\mathbf{x}_f\mathbf{x}_f} \\
&-\dot{\mathbf{U}} = \mathbf{A}'\mathbf{U}, \mathbf{U}(t_f) = L_{\mathbf{x}_f\lambda} \\
&-\dot{\mathbf{W}} = \mathbf{L}'H_{uu}\mathbf{L}, \mathbf{W}(t_f) = 0 \\
&H = F + \mathbf{p}'\mathbf{f}, L = G + \lambda'\mathbf{g}, \mathbf{A} = H_{px} - H_{pu}\mathbf{K} \\
&\mathbf{L} = H_{uu}^{-1}H_{up}\mathbf{U}, \mathbf{K} = H_{uu}^{-1}(H_{ux} + H_{up}\boldsymbol{\Omega}), \mathbf{k} = \mathbf{u}_n + \mathbf{K}\mathbf{x}_n \\
&\mathbf{H} = -\{E_{cc} + \boldsymbol{\Gamma}'[\boldsymbol{\Omega}(0) + \mathbf{U}(0)\mathbf{W}^{-1}(0)\mathbf{U}'(0)]\boldsymbol{\Gamma}\}^{-1}\boldsymbol{\Gamma}'[\boldsymbol{\Omega}(0) \\
&\qquad + \mathbf{U}(0)\mathbf{W}^{-1}(0)\mathbf{U}'(0)]\boldsymbol{\Delta} \\
&\mathbf{M} = \boldsymbol{\Delta} + \boldsymbol{\Gamma}\mathbf{H}, \mathbf{D} = \mathbf{W}^{-1}(0)\mathbf{U}'(0)\mathbf{M}, \lambda = \lambda_n + \mathbf{D}\boldsymbol{\delta}
\end{aligned} \quad (5.23)
$$

In order to impose the asymptotic stability constraint [10] with respect to the nominal state vector $\mathbf{x}_n(t)$, matrices H_{xx}, H_{ux}, and H_{uu} are given by equation (5.19) in terms of parameter ρ.

SIM and SIMS

Design and simulation of linearized optimal control, as outlined above, are implemented by subprogram SIM which is summarized as follows:

SIM function: Design and simulate linearized optimal control for minimizing

$$J = E(\mathbf{c}) + \int_0^{t_f} F(\mathbf{x}, \mathbf{u})dt + G(\mathbf{x}_f)$$

with respect to $\mathbf{u}(t)$ and $\mathbf{c}$ where

$$\dot{\mathbf{x}} = \mathbf{f}(\mathbf{x},\mathbf{u}); \mathbf{x}(0) = \gamma + \boldsymbol{\Gamma}\mathbf{c}$$

and

$$\mathbf{c} = \mathbf{c}_n + \mathbf{H}\boldsymbol{\delta}, \mathbf{u} = \mathbf{k} - \mathbf{J}\begin{bmatrix}\mathbf{x}\\\mathbf{D}\boldsymbol{\delta}\end{bmatrix}.$$

SIM dummy argument list: None.

SIM input data list:

C1/C2/C3/NT,NII,NP,NF/NX,NU,NG,NC,ND/DT,RHO/
GAMM/DELM/$\underbrace{\text{ZV/KV/JM/DELV}}_{\text{NT times}}$

where
  NT = number of trajectory simulations.
  NII = number of integration increments.
  NP = number of integration increments omitted between output.
  NF = number of the trajectory that is used for the solution of backward-time equations.
  $1 \leq NX \leq 7, 1 \leq NU \leq 3, 0 \leq NG \leq 3, 0 \leq NC \leq 1, 1 \leq ND \leq 4$.
  DT = length of integration increments.
  RHO = asymptotic stability parameter (between 0 and 1).
  GAMM(NX,NC) = Γ.
  DELM(NX,ND) = Δ.
  ZV(NX+NG+NC) = vector formed with $x_n(0)$, λ_n, and c_n (on I/O unit 3).
  KV(NU) = k (on I/O unit 3).
  JM(NU,NX+NG) = J (on I/O unit 3).
  DELV(ND) = δ.

SIM output data list: None.

Subprograms called by SIM: MI and SIMS for the computation of
  $E = E(c)$, DEV(NC) = E_c, DDEM(NC,NC) = E_{cc},
  $F = F(y, t)$, FV(NX) = $f(y, t)$, $G = G(x)$,
  DGV(NX) = G_x, DDGM(NX,NX) = G_{xx}, GV(NG) = $g(x)$,
  DGM(NG,NX) = g_x, DDGA(NG,NX,NX) = $(g_1)_{xx}/(g_2)_{xx}/\ldots$,
  DFV(NX+NU) = F_y, DDFM(NX+NU,NX+NU) = F_{yy},
  DFM(NX,NX+NU) = f_y, and DDFA(NX,NX+NU,NX+NU)
  = $(f_1)_{yy}/(f_2)_{yy}/\ldots$, where

$$y = \begin{bmatrix} x \\ u \end{bmatrix}.$$

Subprogram SIM is listed in Table 5-8. User problems are specified for subprogram SIM by writing an appropriate subprogram SIMS. Specifications for writing this subprogram are given as follows:

SIMS function: Compute $E(c), E_c, E_{cc}, F(y, t), f(y, t), G(x), G_x, G_{xx}, g(x),$ $g_x, (g_1)_{xx}/(g_2)_{xx}/\ldots, F_y, F_{yy}, f_y,$ and $(f_1)_{yy}/(f_2)_{yy}/\ldots$ for use with SIM, where

$$y = \begin{bmatrix} x \\ u \end{bmatrix}.$$

Table 5-8
Listing of SIM

```fortran
      SUBROUTINE SIM
C     REAL*8 SLA(NU,NL,NTAB),DDGA(NG,NX,NX),DDFA(NX,NY,NY),
C    1SKM(NU,NTAB),SYM(NY,NTAB),DELM(NX,ND),DFM(NX,NY),DGM(NG,NX),
C    2DDEM(NC,NC),DDFM(NY,NY),DDGM(NX,NX),GAMM(NX,NC),AM(NX,NX),
C    3LM(NU,NL),TLM(NU,NL),HM(MH,MH),PM(NL,NL),DPM(NL,NL),MM(NX,ND),
C    4SDPM(NL,NL),CV(NC),YV(NY),ZV(NZ),FV(NX),GV(NG),DEV(NC),DFV(NY),
C    5DGV(NX),KV(NU),DELV(ND),PV(NX),SFV(NX),SZV(NZ),XV(NL),DPV(NX)
C    6LV(NG)
      REAL*8 SLA(3,10,401),DDGA(4,6,6),DDFA(6,9,9),SKM(3,401),
     1SYM(9,401),DELM(6,4),DFM(6,9),DGM(4,6),DDEM(1,1),
     2DDFM(9,9),DDGM(6,6),GAMM(6,1),AM(6,6),LM(3,10),TLM(3,10),
     3HM(4,4),PM(10,10),DPM(10,10),MM(6,4),SDPM(10,10),CV(1),YV(9),
     4ZV(11),FV(6),GV(4),DEV(1),DFV(9),DGV(6),KV(3),DELV(4),PV(6),
     5SFV(6),SZV(11),XV(10),DPV(6),LV(4)
      REAL*8 B,C,D,E,F,G,G1,G2,G3,G4,G5,G6,G7,G8,DT,DTH,T,RHO,
     1DABS,DSQRT
      DATA MH/4/,MC/1/,MG/4/,MX/6/,MY/9/
      DATA E/0D0/,DEV/0D0/,DDEM/0D0/,F/0D0/,FV/6*0D0/,G/0D0/,
     1DGV/6*0D0/,DDGM/36*0D0/,GV/4*0D0/,DGM/24*0D0/,DDGA/144*0D0/,
     2DFV/9*0D0/,DDFM/81*0D0/,DFM/54*0D0/,DDFA/486*0D0/
      LOGICAL*1 COM(65)
1000  FORMAT(65A1)
1001  FORMAT(5I10)
1002  FORMAT(4D18.10)
1003  FORMAT(4X,65A1)
```

```
26      1004 FORMAT(4X,4D18.10)
27      1005 FORMAT(' 1')
28      1006 FORMAT('      NT,NII,NP,NF')
29      1007 FORMAT('      NX,NU,NG,NC,ND')
30      1008 FORMAT('      DT,RHO')
31      1009 FORMAT('      DELM')
32      1010 FORMAT(///,'   HM')
33      1011 FORMAT(///,'   MM')
34      1012 FORMAT(///,'   ZV')
35      1013 FORMAT(///,'   T,YV,J')
36      1014 FORMAT(///,'   T,JM,KV')
37      1015 FORMAT('      GAMM')
38      1016 FORMAT(///,'   GV')
39      1017 FORMAT(///,'   TRAJECTORY=',I2,'  L=',D18.10)
40      1018 FORMAT(///,'   DELV')
41           WRITE (6,1005)
42           DO 1019 I=1,3
43           READ (5,1000) COM
44      1019 WRITE (6,1003) COM
45           READ (5,1001) NI,NII,NP,NF
46           READ (5,1001) NX,NU,NG,NC,ND
47           NL=NX+NG
48           NZ=NL+NC
49           NY=NX+NU
50           NTAB=NII+1
51           NIC=1
52           READ (5,1002) DT,RHO
```

(continued)

Table 5-8 Continued

```
53           WRITE (6,1006)
54           WRITE (6,1001)    NI,NII,NP,NF
55           WRITE (6,1007)
56           WRITE (6,1001)    NX,NU,NG,NC,ND
57           WRITE (6,1008)
58           WRITE (6,1004)    DT,RHO
59           IF (NG .LE. 0)    GO TO 1021
60           DO 1020 I=1,NG
61           DO 1020 J=1,ND
62  1020     PM(I,J)=0D0
63  1021     IF (NC) 1024,1024,1022
64  1022     READ (5,1002) ((GAMM(I,J),I=1,NX),J=1,NC)
65           WRITE (6,1015)
66           WRITE (6,1004) ((GAMM(I,J),I=1,NX),J=1,NC)
67           DO 1023 I=1,NC
68           DO 1023 J=1,ND
69  1023     HM(I,J)=0D0
70  1024     READ (5,1002) ((DELM(I,J),I=1,NX),J=1,ND)
71           WRITE (6,1009)
72           WRITE (6,1004) ((DELM(I,J),I=1,NX),J=1,ND)
73           DO 1025 I=1,NX
74           DO 1025 J=1,ND
75  1025     MM(I,J)=DELM(I,J)
76           READ (3) (SZV(I),I=1,NZ),((SKM(I,K),I=1,NU),((SLA(I,J,K),
77          1I=1,NU),J=1,NX),K=1,NTAB)
78           REWIND 3
```

```
           IF (NG)       1028,1028,1026
1026       DO 1027   K=1,NTAB
           DO 1027   I=1,NU
           DO 1027   J=1,NG
1027       SLA(I,J+NX,K)=0D0
1028       D=DSQRT(5D-1)
           G1=DT*(1D0-D)
           G2=DT*(1D0+D)
           D=3D0*D
           G6=D-2D0
           G8=-(D+2D0)
           G3=DT/6D0
           G4=-DT/3D0
           D=DSQRT(2D0)
           G5=2D0-D
           G7=2D0+D
           DTH=5D-1*DT
           CALL SIMS(MC,MG,MX,MY,CV,YV,T,E,DEV,DDEM,F,FV,G,DGV,DDGM,
          1GV,DGM,DDGA,DFV,DDFM,DFM,DDFA,1)
1029       IF (NC)   1031,1031,1030
1030       WRITE (6,1010)
           WRITE (6,1004)   ((HM(I,J),I=1,NC),J=1,ND)
1031       WRITE (6,1011)
           WRITE (6,1004)   ((MM(I,J),I=1,NX),J=1,ND)
           WRITE (6,1014)
           NIP=NP
           L=0
           K=NII
```

(continued)

Table 5-8 Continued

```
107        1032 NIP=NIP+1
108             IF (NIP .LE. NP) GO TO 1034
109             T=DT*K
110             WRITE (6,1004) T,((SLA(I,J,K+1),I=1,NU),J=1,NL),
111            1(SKM(I,K+1),I=1,NU)
112             IF (L) 1033,1033,1036
113        1033 NIP=0
114        1034 K=K-1
115             IF (K) 1035,1035,1032
116        1035 L=1
117             NIP=NP
118             GO TO 1032
119        1036 NIP=NP
120             NIIC=0
121             L=0
122             MGILL=1
123             T=0D0
124             READ (5,1002) (DELV(I),I=1,ND)
125             IF (NG) 1040,1040,1037
126        1037 DO 1039 I=1,NG
127             D=0D0
128             DO 1038 J=1,ND
129        1038 D=D+PM(I,J)*DELV(J)
130             IV(I)=D
131        1039 ZV(I+NX)=SZV(I+NX)+D
132        1040 IF (NC) 1044,1044,1041
```

```
133   1041  DO 1043 I=1,NC
134         D=SZV(I+NL)
135         DO 1042 J=1,ND
136   1042  D=D+HM(I,J)*DELV(J)
137         ZV(I+NL)=D
138   1043  CV(I)=D
139         CALL SIMS(MC,MG,MX,MY,CV,YV,T,E,DEV,DDEM,F,FV,G,DGV,DDGM,
140        1GV,DGM,DDGA,DFV,DDFM,DFM,DDFA,2)
141         DO 1046 I=1,NX
142         D=SZV(I)
143         DO 1045 J=1,ND
144   1045  D=D+MM(I,J)*DELV(J)
145         ZV(I)=D
146   1046  XV(I)=D
147         IF (NG) 1049,1049,1047
148   1047  DO 1048 I=1,NG
149   1048  XV(I+NX)=ZV(I+NX)
150   1049  C=E
151         DO 1050 I=1,NU
152         KV(I)=SKM(I,1)
153         DO 1050 J=1,NL
154   1050  LM(I,J)=SLA(I,J,1)
155         WRITE (6,1018)
156         WRITE (6,1004)     (DELV(I),I=1,ND)
157         WRITE (6,1012)
158         WRITE (6,1004)     (ZV(I),I=1,NZ)
159         WRITE (6,1013)
160   1051  DO 1052 I=1,NX
```

(continued)

Table 5-8 Continued

```
161        1052 YV(I)=XV(I)
162             DO 1057 I=1,NU
163             D=KV(I)
164             IF (NG) 1055,1055,1053
165        1053 DO 1054 J=1,NG
166        1054 D=D-LM(I,J+NX)*LV(J)
167        1055 DO 1056 J=1,NX
168        1056 D=D-LM(I,J)*XV(J)
169        1057 YV(I+NX)=D
170             CALL SIMS(MC,MG,MX,MY,CV,YV,T,E,DEV,DDEM,F,FV,G,DGV,DDGM,
171            1GV,DGM,DDGA,DFV,DDFM,DFM,DDFA,3)
172             GO TO (1058,1064,1066,1069),MGILL
173        1058 NIIC=NIIC+1
174             DO 1059 I=1,NY
175        1059 SYM(I,NIIC)=YV(I)
176             NIP=NIP+1
177             IF (NIP .LE. NP) GO TO 1061
178             IF (L) 1060,1060,1071
179        1060 WRITE (6,1004) T,(YV(I),I=1,NY),C
180             NIP=0
181        1061 DO 1062 I=1,NX
182             SFV(I)=FV(I)
183        1062 XV(I)=XV(I)+DTH*FV(I)
184             B=F
185             C=C+DTH*F
186             T=T+DTH
```

```
187         M=NIIC+1
188         DO 1063  I=1,NU
189         KV(I)=5D-1*(SKM(I,NIIC)+SKM(I,M))
190         DO 1063  J=1,NL
191   1063  LM(I,J)=5D-1*(SLA(I,J,NIIC)+SLA(I,J,M))
192         MGILL=2
193         GO TO 1051
194   1064  DO 1065  I=1,NX
195         XV(I)=XV(I)+G1*(FV(I)-SFV(I))
196   1065  SFV(I)=G5*FV(I)+G6*SFV(I)
197         C=C+G1*(F-B)
198         B=G5*F+G6*B
199         MGILL=3
200         GO TO 1051
201   1066  DO 1067  I=1,NX
202         XV(I)=XV(I)+G2*(FV(I)-SFV(I))
203   1067  SFV(I)=G7*FV(I)+G8*SFV(I)
204         C=C+G2*(F-B)
205         B=G7*F+G8*B
206         T=T+DTH
207         DO 1068  I=1,NU
208         KV(I)=SKM(I,M)
209         DO 1068  J=1,NL
210   1068  LM(I,J)=SLA(I,J,M)
211         MGILL=4
212         GO TO 1051
213   1069  DO 1070  I=1,NX
214   1070  XV(I)=XV(I)+G3*FV(I)+G4*SFV(I)
```

(continued)

Table 5-8 Continued

```
215            C=C+G3*F+G4*B
216            MGILL=1
217            IF (NIIC .LT. NII) GO TO 1051
218            I=1
219            NIP=NP
220            GO TO 1051
221       1071 CALL SIMS(MC,MG,MX,MY,CV,YV,T,E,DEV,DDEM,F,FV,G,DGV,DDGM,
222            1GV,DGM,DDGA,DFV,DDFM,DFM,DDFA,4)
223            C=C+G
224            WRITE (6,1004) T,(YV(I),I=1,NY),C
225            IF (NG) 1074,1074,1072
226       1072 WRITE (6,1016)
227            WRITE (6,1004) (GV(I),I=1,NG)
228            DO 1073 I=1,NG
229       1073 C=C+ZV(I+NX)*GV(I)
230       1074 WRITE (6,1017) NIC,C
231            IF (NIC .EQ. NF) GO TO 1075
232            IF (NIC .GE. NI) GO TO 1157
233            NIC=NIC+1
234            GO TO 1036
235       1075 NIC=NIC+1
236            L=0
237            DO 1081 I=1,NX
238            D=DGV(I)
239            IF (NG) 1078,1078,1076
240       1076 DO 1077 J=1,NG
```

```
241  1077  D=D+ZV(J+NX)*DGM(J,I)
242  1078  PV(I)=D
243        DO 1081  J=I,NX
244        D=0D0
245        IF (NG)   1081,1081,1C79
246  1079  DO 1080  K=1,NG
247  1080  D=D+ZV(K+NX)*DDGA(K,I,J)
248  1081  PM(I,J)=DDGM(I,J)+RHO*D
249        IF (NG)   1085,1085,1082
250  1082  DO 1084  J=1,NG
251        DO 1083  I=1,NX
252  1083  PM(I,J+NX)=DGM(J,I)
253        DO 1084  I=1,J
254  1084  PM(I+NX,J+NX)=0D0
255  1085  CALL SIMS(MC,MG,MX,MY,CV,YV,T,E,DEV,DDEM,F,FV,G,DGV,DDGM,
256       1GV,DGM,DDGA,DFV,DDFM,DFM,DDFA,5)
257  1086  DO 1087  I=1,NX
258        DO 1087  J=I,NX
259  1087  PM(J,I)=PM(I,J)
260        DO 1093  I=1,NU
261        DO 1089  J=1,NX
262        C=0D0
263        B=DDFM(J,I+NX)
264        DO 1088  K=1,NX
265        C=C+PV(K)*DDFA(K,J,I+NX)
266  1088  B=B+DFM(K,I+NX)*PM(K,J)
267  1089  TLM(I,J)=B+RHO*C
```

(continued)

Table 5-8 Continued

```
268          IF (NG) 1093,1093,1090
269   1090 DO 1092 J=1,NG
270        C=0D0
271        DO 1091 K=1,NX
272   1091 C=C+DFM(K,I+NX)*PM(K,J+NX)
273   1092 TLM(I,J+NX)=C
274   1093 CONTINUE
275        DO 1095 I=1,NU
276        DO 1095 J=I,NU
277        D=0D0
278        DO 1094 K=1,NX
279   1094 D=D+PV(K)*DDFA(K,I+NX,J+NX)
280        HM(I,J)=DDFM(I+NX,J+NX)+RHO*D
281   1095 HM(J,I)=HM(I,J)
282        CALL MI(MH,NU,HM)
283        DO 1097 I=1,NU
284        DO 1097 J=1,NL
285        D=0D0
286        DO 1096 K=1,NU
287        D=D+HM(I,K)*TLM(K,J)
288   1097 LM(I,J)=D
289        DO 1100 I=1,NX
290        C=DFV(I)
291        DO 1098 J=1,NU
292   1098 C=C-LM(J,I)*DFV(J+NX)
293        DPV(I)=D+C
```

```
294            DO 1100 J=1,NX
295            D=DFM(I,J)
296            DO 1099 K=1,NU
297  1099      D=D-DFM(I,K+NX)*LM(K,J)
298  1100      AM(I,J)=D
299            DO 1104 I=1,NX
300            D=DPV(I)
301            DO 1101 J=1,NX
302  1101      D=D+AM(J,I)*PV(J)
303            DPV(I)=D
304            DO 1104 J=I,NX
305            D=0D0
306            C=DDFM(I,J)
307            DO 1102 K=1,NX
308            D=D+PV(K)*DDFA(K,I,J)
309  1102      C=C+PM(I,K)*DFM(K,J)+DFM(K,I)*PM(K,J)
310            DO 1103 K=1,NU
311  1103      C=C-LM(K,I)*TLM(K,J)
312  1104      DPM(I,J)=C+RHO*D
313            IF (NG) 1110,1110,1105
314  1105      DO 1109 I=1,NG
315            DO 1107 J=1,NX
316            D=0D0
317            DO 1106 K=1,NX
318  1106      D=D+AM(K,J)*PM(K,I+NX)
319  1107      DPM(J,I+NX)=D
320            DO 1109 J=I,NG
321            D=0D0
```

(continued)

Table 5-8 Continued

```
322              DO 1108 K=1,NU
323    1108   D=D+LM(K,I+NX)*TLM(K,J+NX)
324    1109   DPM(I+NX,J+NX)=D
325           GO TO (1111,1119,1122,1127),MGILL
326    1111   DO 1114 I=1,NU
327           D=YV(I+NX)
328           DO 1112 J=1,NX
329    1112   D=D+LM(I,J)*XV(J)
330           DO 1113 J=1,NL
331    1113   SLA(I,J,NIIC)=LM(I,J)
332    1114   SKM(I,NIIC)=D
333           IF (L) 1115,1115,1130
334    1115   DO 1116 I=1,NX
335           SFV(I)=DPV(I)
336    1116   PV(I)=PV(I)+DTH*DPV(I)
337           DO 1117 I=1,NL
338           DO 1117 J=I,NL
339           SDPM(I,J)=DPM(I,J)
340    1117   PM(I,J)=PM(I,J)+DTH*DPM(I,J)
341           T=T-DTH
342           M=NIIC-1
343           DO 1118 I=1,NY
344    1118   YV(I)=5D-1*(SYM(I,M)+SYM(I,NIIC))
345           NIIC=M
346           MGILL=2
347           GO TO 1085
```

```
348        1119  DO 1120 I=1,NX
349              PV(I)=PV(I)+G1*(DPV(I)-SFV(I))
350        1120  SFV(I)=G5*DPV(I)+G6*SFV(I)
351              DO 1121 I=1,NL
352              DO 1121 J=I,NL
353              PM(I,J)=PM(I,J)+G1*(DPM(I,J)-SDPM(I,J))
354        1121  SDPM(I,J)=G5*DPM(I,J)+G6*SDPM(I,J)
355              MGILL=3
356              GO TO 1086
357        1122  DO 1123 I=1,NX
358              PV(I)=PV(I)+G2*(DPV(I)-SFV(I))
359        1123  SFV(I)=G7*DPV(I)+G8*SFV(I)
360              DO 1124 I=1,NL
361              DO 1124 J=I,NL
362              PM(I,J)=PM(I,J)+G2*(DPM(I,J)-SDPM(I,J))
363        1124  SDPM(I,J)=G7*DPM(I,J)+G8*SDPM(I,J)
364              T=T-DTH
365              DO 1125 I=1,NY
366        1125  YV(I)=SYM(I,NIIC)
367              DO 1126 I=1,NX
368        1126  XV(I)=YV(I)
369              MGILL=4
370              GO TO 1085
371        1127  DO 1128 I=1,NX
372        1128  PV(I)=PV(I)+G3*DPV(I)+G4*SFV(I)
373              DO 1129 I=1,NL
374              DO 1129 J=I,NL
375        1129  PM(I,J)=PM(I,J)+G3*DPM(I,J)+G4*SDPM(I,J)
```

(continued)

Table 5-8 Continued

```
376            MGILL=1
377            IF (NIIC .GT. 1) GO TO 1086
378            L=1
379            GO TO 1086
380   1130     IF (NG) 1135,1135,1131
381   1131     DO 1132 I=1,NG
382            DO 1132 J=I,NG
383            HM(I,J)=PM(I+NX,J+NX)
384   1132     HM(J,I)=HM(I,J)
385            CALL MI(MH,NG,HM)
386            DO 1134 I=1,NG
387            DO 1134 J=1,NX
388            D=0D0
389            DO 1133 K=1,NG
390   1133     D=D+HM(I,K)*PM(J,K+NX)
391   1134     DPM(I,J)=D
392   1135     IF (NC) 1149,1149,1136
393   1136     DO 1140 I=1,NX
394            DO 1140 J=I,NX
395            D=PM(I,J)
396            IF (NG) 1139,1139,1137
397   1137     DO 1138 K=1,NG
398   1138     D=D+PM(I,NX+K)*DPM(K,J)
399   1139     SDPM(I,J)=D
400   1140     SDPM(J,I)=D
401            DO 1142 I=1,NC
```

```
402           DO 1142 J=1,NX
403           D=0D0
404           DO 1141 K=1,NX
405 1141      D=D+GAMM(K,I)*SDPM(K,J)
406 1142      PM(I,J)=D
407           DO 1144 I=1,NC
408           DO 1144 J=I,NC
409           D=DDEM(I,J)
410           DO 1143 K=1,NX
411 1143      D=D+PM(I,K)*GAMM(K,J)
412           HM(I,J)=D
413 1144      HM(J,I)=D
414           CALL MI(MH,NC,HM)
415           DO 1146 I=1,NC
416           DO 1146 J=1,NX
417           D=0D0
418           DO 1145 K=1,NC
419 1145      D=D+HM(I,K)*PM(K,J)
420 1146      SDPM(I,J)=D
421           DO 1148 I=1,NC
422           DO 1148 J=1,ND
423           D=0D0
424           DO 1147 K=1,NX
425 1147      D=D-SDPM(I,K)*DELM(K,J)
426 1148      HM(I,J)=D
427           DO 1152 I=1,NX
428           DO 1152 J=1,ND
```

(continued)

Table 5-8 Continued

```
429              D=DELM(I,J)
430              IF (NC) 1152,1152,1150
431         1150 DO 1151 K=1,NC
432         1151 D=D+GAMM(I,K)*HM(K,J)
433         1152 MM(I,J)=D
434              IF (NG) 1156,1156,1153
435         1153 DO 1155 I=1,NG
436              DO 1155 J=1,ND
437              D=0D0
438              DO 1154 K=1,NX
439         1154 D=D+DPM(I,K)*MM(K,J)
440         1155 PM(I,J)=D
441         1156 IF (NIC .LE. NI) GO TO 1029
442         1157 WRITE (6,1005)
443              RETURN
444              END
```

SIMS argument list: MC,MG,MX,MY,CV,YV,T,E,DEV,DDEM,F,FV,G,DGV, DDGM,GV,DGM,DDGA,DFV,DDFM,DFM,DDFA,MODE, where
    CV(MC) = c.
    YV(MY) = y.
    T = t.
    E = $E(\mathbf{c})$.
    DEV(MC) = $E_\mathbf{c}$.
    DDEM(MC,MC) = $E_{\mathbf{cc}}$.
    F = $F(\mathbf{y}, t)$.
    FV(MX) = $\mathbf{f}(\mathbf{y}, t)$.
    G = $G(\mathbf{x})$.
    DGV(MX) = $G_\mathbf{x}$.
    DDGM(MX,MX) = $G_{\mathbf{xx}}$.
    GV(MG) = $\mathbf{g}(\mathbf{x})$.
    DGM(MG,MX) = $\mathbf{g}_\mathbf{x}$.
    DDGA(MG,MX,MX) = $(g_1)_{\mathbf{xx}}/(g_2)_{\mathbf{xx}}/\ldots$
    DFV(MY) = $F_\mathbf{y}$.
    DDFM(MY,MY) = $F_{\mathbf{yy}}$.
    DFM(MX,MY) = $\mathbf{f}_\mathbf{y}$.
    DDFA(MX,MY,MY) = $(f_1)_{\mathbf{yy}}/(f_2)_{\mathbf{yy}}/\ldots$
    MODE = 1 for program initialization, 2 for the computation of E, DEV, and DDEM, 3 for the computation of F and FV, 4 for the computation of G, DGV, DDGM, GV, DGM, and DDGA, and 5 for the computation of DFV, DDFM, DFM, and DDFA.

Example 8B

An example of the use of subprogram SIM is provided by the problem stated in Example 8A. Calling program SIMC and subprogram SIMS are listed in Table 5-9a for this example. The user should note that the statements in subprograms MIN5S and SIMS can be identical if one wishes to eliminate the task of reprogramming for SIMS. A listing of program input cards for this example is given in Table 5-9b. Finally, program printed output is listed in Table 5-9c for this example. The first two trajectories shown illustrate the consequences of perturbations in initial state when the parameter vector c and Lagrange multiplier vector $\boldsymbol{\lambda}$ are not adjusted to account for the perturbation. The third trajectory, corresponding to condition NG = 3, is used as the nominal trajectory for backward-time computations. This trajectory is also optimal because condition $\boldsymbol{\delta} = \mathbf{0}$ also holds. The final three trajectories are a repeat of the first three except that vectors c and $\boldsymbol{\lambda}$ are adjusted as linear functions of perturbation vector $\boldsymbol{\delta}$. Linearized adjustment of vectors c and $\boldsymbol{\lambda}$ decreased terminal errors by two orders of magnitude for this example.

Table 5-9a
Listing of User-supplied Programs for Example 8B

```
 1     C     SIMC EXAMPLE 8B
 2           CALL SIM
 3           STOP
 4           END
 5           SUBROUTINE SIMS(MC,MG,MX,MY,CV,YV,T,E,DEV,
 6          1DDEM,F,FV,G,DGV,DDGM,GV,DGM,DDGA,DFV,DDFM,
 7          2DFM,DDFA,MODE)
 8           REAL*8 DDGA(MG,MX,MX),DDFA(MX,MY,MY),DDEM(MC,MC),
 9          1DDGM(MX,MX),DGM(MG,MX),DDFM(MY,MY),DFM(MX,MY),
10          2CV(MC),YV(MY),DEV(MC),FV(MX),DGV(MX),GV(MG),
11          3DFV(MY),T,E,F,G
12           REAL*8 TF
13    1000  FORMAT (D18.10)
14    1001  FORMAT (//,'          TF=',D18.10)
15           GO TO (1002,1003,1004,1005,1006),MODE
16    1002  READ (5,1000) TF
17           WRITE (6,1001) TF
18           DDFA(1,2,3)=1.D0
19           DGM(1,1)=1.D0
20           DDGM(1,1)=1.D0
21           DEV(1)=TF
22           RETURN
23    1003  E=CV(1)*TF
24           RETURN
```

```
25  1004  FV(1)=YV(2)*YV(3)
26        F=FV(1)*YV(3)
27        RETURN
28  1005  GV(1)=YV(1)
29        DGV(1)=YV(1)
30        G=.5D0*YV(1)**2
31        RETURN
32  1006  DFV(2)=YV(3)**2
33        DDFM(2,3)=2.D0*YV(3)
34        DDFM(3,3)=2.D0*YV(2)
35        DFV(3)=DDFM(3,3)*YV(3)
36        DFM(1,2)=YV(3)
37        DFM(1,3)=YV(2)
38        RETURN
39        END
```

Table 5-9b
Listing of User-supplied Input Data for Example 8B

```
 1    SIM(MI,SIMS)    EXAMPLE 8B
 2    MERRIAM
 3    UNIVERSITY OF ROCHESTER
 4               6          50           9           3
 5               2           1           1           1
 6        .02               1.
 7        0.                1.
 8        1.                0.
 9        1.
10    -.05
11     .05
12     0.
13    -.05
14     .05
15     0.
```

Table 5-9c
Printed Output from SIM for Example 8B

```
SIM(MI,SIMS)   EXAMPLE 8B
MERRIAM
UNIVERSITY OF ROCHESTER
NT,NII,NP,NF
   6    50     9     3
NX,NU,NG,NC,ND
   2     1     1     1
DT,RHO
  0.2000000000D-01  0.1000000000D 01
GAMM
  0.0000000000D 00  0.1000000000D 01
DELM
  0.1000000000D 01  0.0000000000D 00

TF=  0.1000000000D 01

HM
  0.0000000000D 00

MM
  0.1000000000D 01  0.0000000000D 00
```

(continued)

Table 5-9c Continued

```
T,JM,KV
 0.1000000000D 01    0.5000000000D 00    0.2424846428D-04    0.0000000000D 00
-0.9999757510D 00    0.4545451919D 00   -0.9088205796D-01    0.0000000000D 00
-0.8000000000D 00    0.4166662252D 00   -0.1666369997D 00    0.0000000000D 00
-0.9999716473D 00    0.3846148204D 00   -0.2307370841D 00    0.0000000000D 00
-0.6000000000D 00    0.3571422085D 00   -0.2856798077D 00    0.0000000000D 00
-0.9999682030D 00    0.3333326270D 00   -0.3332966623D 00    0.0000000000D 00
-0.4000000000D 00
-0.9999652431D 00
-0.2000000000D 00
-0.9999626501D 00
 0.0000000000D 00
-0.9999603423D 00

DELV
-0.5000000000D-01

ZV
 0.9500000000D 00    0.9999889230D 00    0.1999999999D 01    0.9999889230D 00

T,YV,J
 0.0000000000D 00    0.9500000000D 00    0.9999889230D 00   -0.9833333676D 00
 0.9999889230D 00
```

0.20000000000D 00	0.75333339453D 00	0.99998892300D 00	-0.98333335590D 00
0.11933780750D 01	0.55666765280D 00	0.99998892300D 00	-0.98333334420D 00
0.40000000000D 00	0.36000106250D 00	0.99998892300D 00	-0.98333333300D 00
0.13867690460D 01		
0.60000000000D 00	0.16333409230D 00	0.99998892300D 00	-0.98333332240D 00
0.15801599270D 01		
0.80000000000D 00	-0.33333372480D-01	0.99998892300D 00	-0.98333331290D 00
0.17735515550D 01		
0.10000000000D 01		
0.19674997120D 01		

GV
-0.33333372480D-01

TRAJECTORY= 1   L= 0.19008329670D 01

DELV
 0.50000000000D-01

ZV
 0.10500000000D 01   0.99998892300D 00   0.19999999990D 01   0.99998892300D 00

(continued)

Table 5-9c Continued

```
T,YV,J
0.000000000D 00    0.1050000000D 01    0.999989230D 00   -0.1016666630D 01
0.999989230D 00
0.200000000D 00    0.8466673115D 00    0.999989230D 00   -0.1016666640D 01
0.1206712124D 01
0.400000000D 00    0.6433343729D 00    0.999989230D 00   -0.1016666649D 01
0.1413435833D 01
0.600000000D 00    0.4400011223D 00    0.999989230D 00   -0.1016666656D 01
0.1620160177D 01
0.800000000D 00    0.2366674757D 00    0.999989230D 00   -0.1016666659D 01
0.1826885326D 01
0.1000000000D 01    0.3333331555D-01    0.999989230D 00   -0.1016666657D 01
0.2034167075D 01
```

GV
  0.3333331555D-01

TRAJECTORY= 2   L=   0.2100833706D 01

DELV
  0.0000000000D 00

ZV
 0.1000000000D 01 0.9999889230D 00 0.1999999999D 01 0.1999999999D 01

T,YV,J
 0.0000000000D 00 0.1000000000D 01 0.9999889230D 00 0.9999889230D 00
 0.9999889230D 00 0.8000006284D 00 0.9999889230D 00 -0.9999999990D 00
 0.2000000000D 00 0.6000010129D 00 0.9999889230D 00 -0.9999999981D 00
 0.1199989882D 01 0.4000010924D 00 0.9999889230D 00 -0.9999999967D 00
 0.4000000000D 00 0.2000007840D 00 0.9999889230D 00 -0.9999999944D 00
 0.1399991328D 01 -0.2846336605D-07 0.9999889230D 00 -0.9999999908D 00
 0.6000000000D 00   -0.9999999849D 00
 0.1599993384D 01
 0.8000000000D 00
 0.1799996217D 01
 0.1000000000D 01
 0.2000000057D 01

GV
 -0.2846336605D-07

TRAJECTORY= 3   L=   0.2000000000D 01

HM
  0.9222531461D 00

(continued)

Table 5-9c Continued

```
MM
  0.1000000000D 01   0.9222531461D 00

T,JM,KV
  0.1000000000D 01                      0.5000000000D 00   0.5000000000D 00
 -0.9999999987D 00                      0.4545459123D 00  -0.4758583165D-01   0.4545459123D 00
 -0.9566757565D 00                      0.4166674359D 00  -0.9069861085D-01   0.4166674359D 00
 -0.6000000000D 00                      0.3846163678D 00  -0.1298589843D 00   0.3846163678D 00
 -0.9240301711D 00                      0.3571439875D 00  -0.1655446283D 00   0.3571439875D 00
 -0.4000000000D 00                      0.3333345642D 00  -0.1981772717D 00   0.3333345642D 00
 -0.8990873323D 00
 -0.2000000000D 00
 -0.8798273782D 00
  0.0000000000D 00
 -0.8648405113D 00

DELV
 -0.5000000000D-01

ZV
  0.9500000000D 00   0.9538762657D 00   0.2000943856D 01   0.9538762657D 00
```

T,YV,J

0.0000000000D 00	0.9500000000D 00	0.9538762657D 00
0.9538762657D 00		-0.9927863715D 00
0.2000000000D 00	0.7604991918D 00	0.9538762657D 00
0.1142111161D 01		-0.9938630930D 00
0.4000000000D 00	0.5707769905D 00	0.9538762657D 00
0.1330786148D 01		-0.9951111250D 00
0.6000000000D 00	0.3807967285D 00	0.9538762657D 00
0.1519974766D 01		-0.9965737897D 00
0.8000000000D 00	0.1905120379D 00	0.9538762657D 00
0.1709770203D 01		-0.9983102554D 00
0.1000000000D 01	-0.1368809189D-03	0.9538762657D 00
0.1900292951D 01		-0.1000403487D 01

GV
 -0.1368809189D-03

TRAJECTORY= 4   L= 0.1900019060D 01

DELV
 0.5000000000D-01

ZV
 0.1050000000D 01   0.1046101580D 01   0.1999056142D 01   0.1046101580D 01

(continued)

Table 5-9c Continued

```
T,YV,J
 0.00000000D  00    0.10500000D  01    0.10461015 80D  01   -0.10072136 26D  01
 0.10461015 80D  01
 0.20000000D  00    0.83938321 05D  00    0.10461015 80D  01   -0.10060944 55D  01
 0.12581241 86D  01
 0.40000000D  00    0.62901710 68D  00    0.10461015 80D  01   -0.10048088 96D  01
 0.14696423 81D  01
 0.60000000D  00    0.41894052 89D  00    0.10461015 80D  01   -0.10033158 13D  01
 0.16805787 64D  01
 0.80000000D  00    0.20920235 68D  00    0.10461015 80D  01   -0.10015591 93D  01
 0.18908361 27D  01
 0.10000000D  01   -0.13467440 51D-03    0.10461015 80D  01   -0.99946073 36D  00
 0.21002900 24D  01

GV
-0.13467440 51D-03

TRAJECTORY=  5   L=   0.21000208 03D  01

DELV
 0.00000000 00D  00
```

```
ZV
  0.100000000D 01    0.999889230D 00    0.199999999D 01    0.999889230D 00

T,YV,J
  0.000000000D 00    0.100000000D 01    0.999889230D 00   -0.999999999D 00
  0.999889230D 00
  0.200000000D 00    0.800006284D 00    0.999889230D 00   -0.999999981D 00
  0.119989882D 01
  0.400000000D 00    0.600010129D 00    0.999889230D 00   -0.999999967D 00
  0.139999132D 01
  0.600000000D 00    0.400010924D 00    0.999889230D 00   -0.999999945D 00
  0.159999338D 01
  0.800000000D 00    0.200007841D 00    0.999889230D 00   -0.999999908D 00
  0.179999621D 01
  0.100000000D 01   -0.283521628D-07    0.999889230D 00   -0.999999850D 00
  0.200000057D 01

GV
 -0.283521628D-07

TRAJECTORY=  6   L=   0.200000000D 01
```

6 Utility Programs

Almost all of the computer programs described in previous chapters are implemented by calls to one or more utility subprograms, such as a subprogram for matrix inversion. These utility subprograms are summarized here. System-supplied utility subprograms are related to one of two general categories of computational problems.

The first category is factoring polynomials, and the computer programs and associated computational techniques for this category are discussed in some detail. Factoring polynomials is accomplished by a parameter optimization technique, which compares very favorably with the other highly sophisticated techniques presented in the literature.

The second category of system-supplied utility subprograms is associated with solving linear equations. All of the computational techniques implemented in this category have been well documented in the literature, and only very brief summaries of these techniques are given here.

6.1 Factoring Polynomials

Description of the Basic Methods

This section is concerned with computing the factors of a polynomial $p(s)$ which is defined by

$$p(s) = p_1 s^n + p_2 s^{n-1} + \ldots + p_{n+1} \qquad (6.1)$$

where coefficient vector **p** is real and n is referred to as deg $p(s)$. Thus, values of the complex number

$$s = s_1 + j s_2 \qquad (6.2)$$

are sought such that $p(s) = 0$.

This problem can be formulated as a two-dimensional parameter optimization problem by defining an objective function

$$f(s) = \frac{1}{2} p(s) p(s^*), \qquad (6.3)$$

where s^* denotes the complex conjugate of s. Gradient vector f_s and hessian matrix f_{ss} are easily derived for this problem in terms of additional polynomials.

$$q(s) = \frac{dp(s)}{ds} \text{ and } r(s) = \frac{dq(s)}{ds}. \tag{6.4}$$

Properties

$$p(s^*) = p^*(s),$$

and
$$\left.\begin{array}{l}q(s)p(s^*) + p(s)q(s^*) = 2Re\{p(s^*)q(s)\} = 2Re\{p(s)q(s^*)\},\\ q(s)p(s^*) - p(s)q(s^*) = 2jIm\{p(s^*)q(s)\} = -2jIm\{p(s)q(s^*)\}\end{array}\right\} \tag{6.5}$$

are also used where Re and Im denote real and imaginary parts respectively. Appropriate differentiations of the objective function then yield

$$f_s = \begin{bmatrix} Re\{p(s)q(s^*)\} \\ Im\{p(s)q(s^*)\} \end{bmatrix} \tag{6.6}$$

and

$$f_{ss} = \begin{bmatrix} Re\{p(s)r(s^*)\} & Im\{p(s)r(s^*)\} \\ Im\{p(s)r(s^*)\} & -Re\{p(s)r(s^*)\} \end{bmatrix} + \begin{bmatrix} q(s)q(s^*) & 0 \\ 0 & q(s)q(s^*) \end{bmatrix}. \tag{6.7}$$

Matrix f_{cc} is positive definite if and only if

$$|q(s)|^2 > |p(s)r(s)| \tag{6.8}$$

where notation

$$|q(s)| = \sqrt{q(s)q(s^*)} \tag{6.9}$$

is used for the magnitude of a complex number $q(s)$.

Condition $|q(s)| > 0$ whenever $p(s) = 0$ holds, as implied by equation (6.8), requires that the multiplicity of each zero location be no greater than one. Thus, use of the Newton-Raphson method for factoring polynomials requires that only first-order factors appear in the polynomials. Moreover, application of the Newton-Raphson method to this subset of polynomials also requires an initial approximation to a zero location of $p(s)$ of sufficient accuracy so that equation (6.8) holds on each subsequent iteration. These difficulties are typical of applying the Newton-Raphson technique to any optimization problem. Application of the Newton-Raphson method to the problem of factoring polynomials can be accomplished by the following techniques:

The greatest common divisor polynomial of polynomials $p(s)$ and $dp(s)/$

ds, which is denoted as $p(s) \gcd dp(s)/ds$, contains all of the excess factors of $p(s)$. Thus, polynomial $p(s)/(p(s) \gcd dp(s)/ds)$ contains only first-order factors and can be used to specify all distinct zero locations of polynomial $p(s)$. Moreover, a continuation of this process to polynomial $p(s) \gcd dp(s)/ds$ can be used, at least in theory, to determine the multiplicity of the zero locations of polynomial $p(s)$.

Initialization of the Newton-Raphson method, such that equation (6.8) holds for a polynomial $p(s)$ with only first-order factors, can always be accomplished in theory by application of the Lehmer-Schur test [11]. Let an arbitrary point s be expressed as

$$s = rz + c, \qquad (6.10)$$

where parameters r and c are real and complex respectively. Then every circle in the s-plane can be generated using $z = e^{j\theta}$ and $0 \leq \theta \leq 2\pi$ by proper choice of radius r and center c. The Lehmer-Schur test determines whether any zeros of a polynomial are contained in a unit circle which is centered at the origin. Therefore, polynomial $p(s)$ first is rewritten as a polynomial $f(z)$ which is defined as

$$f(z) = p(rz + c) \qquad (6.11)$$

and is expressed with complex coefficients as

$$f(z) = f_1 z^n + f_2 z^{(n-1)} + \ldots + f_{n+1}. \qquad (6.12)$$

Then a second polynomial $g(z)$ is defined as

$$g(z) = z^n f^*(1/z^*) = f^*_{n+1} z^n + f^*_n z^{(n-1)} + \ldots + f^*_1 \qquad (6.13)$$

for the purpose of defining transformation T as

$$T[f(z)] = f_{n+1} f(z) - f^*_1 g(z). \qquad (6.14)$$

Note that condition $\deg T[f(z)] < \deg f(z)$ holds so that the sequence of transformations

$$T^k[f(z)] = T\{T^{k-1}[f(z)]\} \qquad (6.15)$$

for $k = 1, 2, \ldots$ terminates with $k \leq n$. Also note that

$$T[f(0)] = f_{n+1} f^*_{n+1} - f_1 f^*_1 \qquad (6.16)$$

is a real number so that the following result can be stated.

Suppose k is defined as the smallest integer for which $T^k[f(0)] = 0$. Also suppose that condition $k > 0$ holds so that condition $f(0) \neq 0$ also holds. Then two results can be stated as follows:

1. If condition $T^j[f(0)] < 0$ holds for some j with $0 < j < k$, then $f(z)$ has at least one zero inside the unit circle.
2. If condition $T^j[f(0)] > 0$ holds for some j with $1 \leq j < k$ and if $T^{k-1}[f(z)]$ is a constant, then no zero of $f(z)$ lies inside the unit circle.

These results do not apply to the situation where $T^k[f(0)] = 0$ holds but $T^{k-1}[f(z)]$ is not a constant.

Proper initialization of the Newton-Raphson method can always be accomplished by systematic application of the above results. For example, an attempt can be made to locate a zero of polynomial $p(s)$ inside the unit circle closest to the origin. If polynomial $p(s)$ has a zero on the boundary of the unit circle, the above results do not apply, and, thus, the procedure would commence with a circle of radius $r = 3/2$ instead of $r = 1$. If all of the zeros of polynomial $p(s)$ lie outside the unit circle, the procedure can be applied instead to polynomial $s^n p(1/s)$.

Locating a zero of polynomial $p(s)$ inside the unit circle closest to the origin proceeds by applying the above results to a sequence of circles generated by halving their radii. Once result (2) applies to a circle of radius r, then no zero lies in the circle of radius r, but a zero is known to lie in the annulus formed by circles with radii r and $2r$. This annulus can be systematically searched by searching each of the eight circles of radius $4r/5$ and centers given by

$$c + \frac{3r}{2\cos(\pi/8)} e^{j(13-4n)\pi/16}$$

for $n = 0, 1, \ldots, 7$.

This systematic process of selecting a sequence of circles for the application of results (1) and (2) is continued until a circle is found which contains a zero of polynomial $p(s)$ and results in convexity condition

$$|q(c)|^2/|p(c)r(c)| > \textit{ratio} \geq 1. \qquad (6.17)$$

If parameter *ratio* is sufficiently large and radius r is sufficiently small, then the convexity condition given in equation (6.8) can be made to hold for all s inside the circle, and thus the circle contains at most one zero of polynomial $p(s)$. When this condition is achieved, the Newton-Raphson method can be applied readily to locating the zero rapidly and with great accuracy.

Polynomials of high degree cause the coefficients of polynomial $T^k[f(z)]$ to become very large. Results (1) and (2) can be applied to a sequence of normalized polynomials, however, because condition

UTILITY PROGRAMS

$$T[\alpha f(z)] = \alpha\alpha^* T[f(z)] \tag{6.18}$$

holds for any nonzero complex number α.

FACTOR

Methods of reducing polynomial $p(s)$ to a set of polynomials, which have only first-order factors and are not zero at the origin, are implemented in subprogram FACTOR. The computational flow chart for FACTOR is given in Figure 6-1.

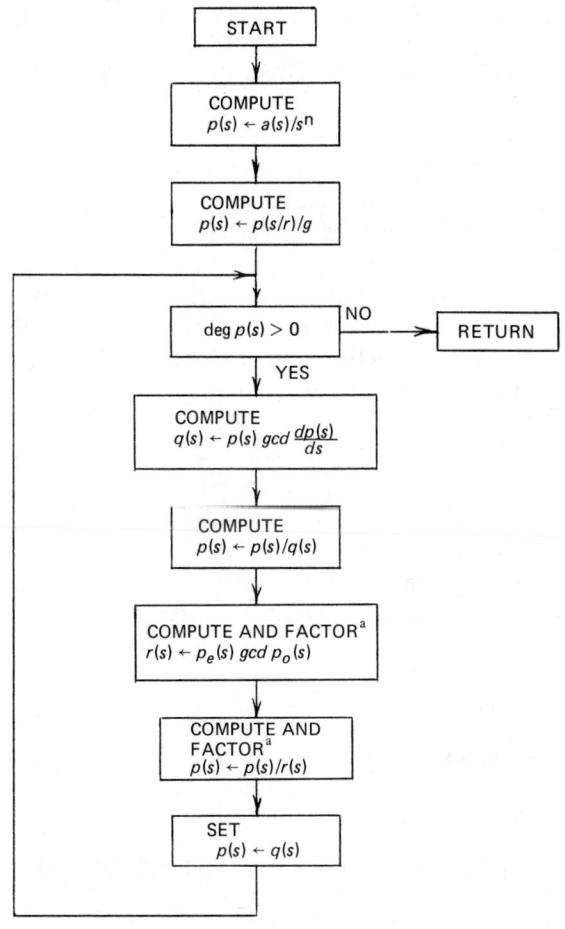

[a] Implemented by calling subprogram FACT.
Figure 6-1. Computational Flow Chart for Subprogram FACTOR

As indicated, the first step in FACTOR is the elimination of zeros at the origin from polynomial $a(s)$. Then the resulting polynomial is normalized by gain g and radius r for the eventual purpose of eliminating excess factors from the polynomial. Finally, a set of polynomials is generated by the application of a method for computing the greatest polynomial divisors for polynomials $p(s)$ and $dp(s)/ds$ and polynomials $p_e(s)$ and $p_o(s)$ where subscripts e and o denote even and odd parts of polynomial $p(s)$ respectively. Each of the polynomials in the set has only first-order factors and is factored by calling subprogram FACT as it is generated. Subprogram FACTOR is summarized as follows:

FACTOR function: Compute the zero locations and their multiplicity for polynomial $a(s)$ as a subprogram.

FACTOR dummy argument list: N,AV,GAIN,RADIUS,EPSLN,PRINT,L,ZM, MV, where
    $1 \leqslant N \leqslant 8$.
    AV(8) = coefficient vector a for polynomial $a(s)$ which has N coefficients.
    GAIN = order of magnitude of the leading coefficient of polynomial $a(s)$ which is used to determine the degree of the polynomial to be factored after zeros at the origin have been removed.
    RADIUS = order of magnitude of the geometric mean of nonzero radii of the zero locations of polynomial $a(s)$ which is used to normalize the polynomial to have a unit geometric mean of zero locations.
    EPSLN = threshold for treating zero locations of the normalized form of polynomial $a(s)$ as distinct.
    PRINT = logical control ("on" for value true and "off" for value false) for whether the steps in locating zeros of polynomial $a(s)$ are to be printed.
    L = number of distinct zero locations, with non-negative imaginary parts, of polynomial $a(s)$ that are found.
    ZM(2,7) = distinct zero locations with non-negative imaginary parts of polynomial $a(s)$ where real and imaginary parts are in rows one and two respectively.
    MV(7) = multiplicity of distinct zero locations with non-negative real parts of polynomial $a(s)$.

FACTOR input data list: None.

FACTOR output data list: None.

Subroutines called by FACTOR: CONVEX,EVAL,FACT,INTL,MATCH,PC,PD, SEARCH.

Subprogram FACTOR is listed in Table 6-1.

UTILITY PROGRAMS 287

Table 6-1
Listing of FACTOR

```
1           SUBROUTINE FACTOR (N,AV,GAIN,RADIUS,EPSLN,PRINT,
2          1                   L,ZM,MV)
3     C    FACTOR ARBITRARY REAL POLYNOMIALS
4           INTEGER*4 MV(N-1)
5           REAL*8 ZM(2,N-1),YM(2,N-1),AV(N),PV(N),QV(N),RV(N),SV(N)
6           INTEGER*4 MV(7)
7           LOGICAL*1 PRINT,MATCH,H
8           REAL*8 ZM(2,7),YM(2,7),AV(8),PV(8),QV(8),
9          1RV(8),SV(8),D,E,GAIN,RADIUS,EPSLN,
10         2EPSLN2,DABS,DSQRT
11    1000 FORMAT (/,4X,'REMAINDER POLY FOR MULTIPLE ZERO DEFLATION')
12    1001 FORMAT (4X,4D18.10)
13    1002 FORMAT (/,4X,'REMAINDER POLY FOR EVEN FACTOR DEFLATION')
14    1003 FORMAT (//,4X,'ROOT ',2D18.10,' OF MULT ',I2,
15         1' NOT SAME TYPE AS ',I2,'TH ROOT')
16          EPSLN2=EPSLN**2
17          M=N
18          DO 1004 I=1,M
19    1004 PV(I)=AV(I)
20          D=GAIN*EPSLN2
21          L=0
22    1005 J=L+1
23          IF (J-M) 1006,1050,1050
24    1006 IF (DABS(PV(J))-D) 1007,1007,1008
25    1007 L=J
```

(continued)

Table 6-1 Continued

```
26            GO TO 1005
27      1008  M=M-L
28            N=M
29            D=PV(J)
30            DO 1009 I=1,M
31            AV(I)=AV(I+L)
32      1009  PV(I)=PV(I+L)/(D*RADIUS**(I-1))
33            I=0
34            LS=1
35            MULT=0
36      1010  IF (DABS(PV(M))-EPSLN2) 1011,1011,1012
37      1011  L=L+1
38            M=M-1
39            GO TO 1010
40      1012  IF (L) 1014,1014,1013
41      1013  ZM(1,1)=0.D0
42            ZM(2,1)=0.D0
43            MV(1)=L
44            L=1
45            LS=2
46      1014  J=M-1
47            MULT=MULT+1
48            IF (J) 1049,1049,1015
49      1015  DO 1016 I=1,J
50      1016  QV(I)=(M-I)*PV(I)
51            CALL PC (M,PV,J,QV,EPSLN,JJ,SV)
```

```
52        CALL PD (M,PV,JJ,SV,0.D0,M,PV,KK,QV)
53        IF (.NOT. PRINT) GO TO 1017
54        WRITE (6,1000)
55        WRITE (6,1001) (QV(I),I=1,KK)
56   1017 IF (M-3) 1041,1018,1018
57   1018 DO 1019 I=2,M,2
58        QV(I)=0.D0
59   1019 RV(I)=0.D0
60        DO 1020 I=1,M,2
61   1020 QV(I)=PV(I)
62        J=M
63        K=J-1
64        DO 1021 I=1,K,2
65   1021 RV(I)=PV(I+1)
66        KK=1
67   1022 IF (KK-K) 1023,1027,1027
68   1023 IF (DABS(RV(KK))-EPSLN2) 1024,1024,1025
69   1024 KK=KK+2
70        GO TO 1022
71   1025 KK=KK-1
72        K=K-KK
73        DO 1026 I=1,K,2
74   1026 RV(I)=RV(I+KK)
75        CALL PC (M,QV,K,RV,EPSLN,J,QV)
76        IF (J-1) 1041,1041,1027
77   1027 CALL PD (M,PV,J,QV,0.D0,M,PV,KK,RV)
78        IF (.NOT. PRINT) GO TO 1028
```

(continued)

Table 6-1 Continued

```
79         WRITE (6,1002)
80         WRITE (6,1001) (RV(I),I=1,KK)
81    1028 K=1+J/2
82         DO 1029 I=2,K
83    1029 QV(I)=QV(2*I-1)
84         CALL FACT (K,QV,EPSLN2,PRINT,LL,YM)
85         DO 1040 I=1,LL
86         IF (YM(2,I)) 1037,1030,1037
87    1030 D=DSQRT(DABS(YM(1,I)))*RADIUS
88         IF (YM(1,I)) 1034,1034,1031
89    1031 IF (MULT-1) 1032,1032,1033
90    1032 L=L+1
91         ZM(1,L)=-D
92         ZM(2,L)=0.D0
93         MV(L)=1
94         L=L+1
95         ZM(1,L)=D
96         ZM(2,L)=0.D0
97         MV(L)=1
98         GO TO 1040
99    1033 D=-D
100        E=0.D0
101        IF (.NOT. MATCH(LS,L,ZM,D,E,K))
102   1WRITE (6,1003) D,E,MULT,K
103        MV(K)=MULT
104        MV(K+1)=MULT
```

```
105              GO TO 1040
106  1034    IF (MULT-1) 1035,1035,1036
107  1035    L=L+1
108          ZM(1,L) = 0.D0
109          ZM(2,L) = D
110          MV(L) = 1
111          GO TO 1040
112  1036    E=0.D0
113          IF (.NOT. MATCH(LS,L,ZM,E,D,K))
114         1WRITE (6,1003) E,D,MULT,K
115          MV(K)=MULT
116          GO TO 1040
117  1037    D=DSQRT(YM(1,I)**2+YM(2,I)**2)
118          E=DSQRT(.5D0*DABS(D-YM(1,I)))*RADIUS
119          D=-DSQRT(.5D0*DABS(D+YM(1,I)))*RADIUS
120          IF (MULT-1) 1038,1038,1039
121  1038    L=L+1
122          ZM(1,L) = D
123          ZM(2,L) = E
124          MV(L) = 1
125          L=L+1
126          ZM(1,L) = -D
127          ZM(2,L) = E
128          MV(L) = 1
129          GO TO 1040
130  1039    IF (.NOT. MATCH(LS,L,ZM,D,E,K))
131         1WRITE (6,1003) D,E,MULT,K
```

(continued)

Table 6-1 Continued

```
132            MV(K)=MULT
133            MV(K+1)=MULT
134      1040  CONTINUE
135      1041  IF (M-1)  1047,1047,1042
136      1042  CALL FACT (M,PV,EPSLN2,PRINT,LL,YM)
137            IF (MULT-1) 1043,1043,1045
138      1043  DO 1044 I=1,LL
139            ZM(1,I+L)=YM(1,I)*RADIUS
140            ZM(2,I+L)=YM(2,I)*RADIUS
141      1044  MV(I+L)=1
142            L=L+LL
143            GO TO 1047
144      1045  DO 1046 I=1,LL
145            D=YM(1,I)*RADIUS
146            E=YM(2,I)*RADIUS
147            IF (.NOT. MATCH(LS,L,ZM,D,E,K))
148           1WRITE (6,1003) D,E,MULT,K
149      1046  MV(K)=MULT
150      1047  M=JJ
151            DO 1048 I=1,M
152      1048  PV(I)=SV(I)
153            GO TO 1014
154      1049  RETURN
155      1050  N=1
156            AV(1)=0.D0
157            RETURN
158            END
```

UTILITY PROGRAMS

FACT

Methods used to factor a polynomial with only first-order factors are implemented in subprogram FACT as indicated in Figure 6-2. The first method employed in FACT is to remove each factor from the polynomial as the corresponding zero location is found. Removal of factors introduces numerical error so a second method is also included; namely, each zero location is refined

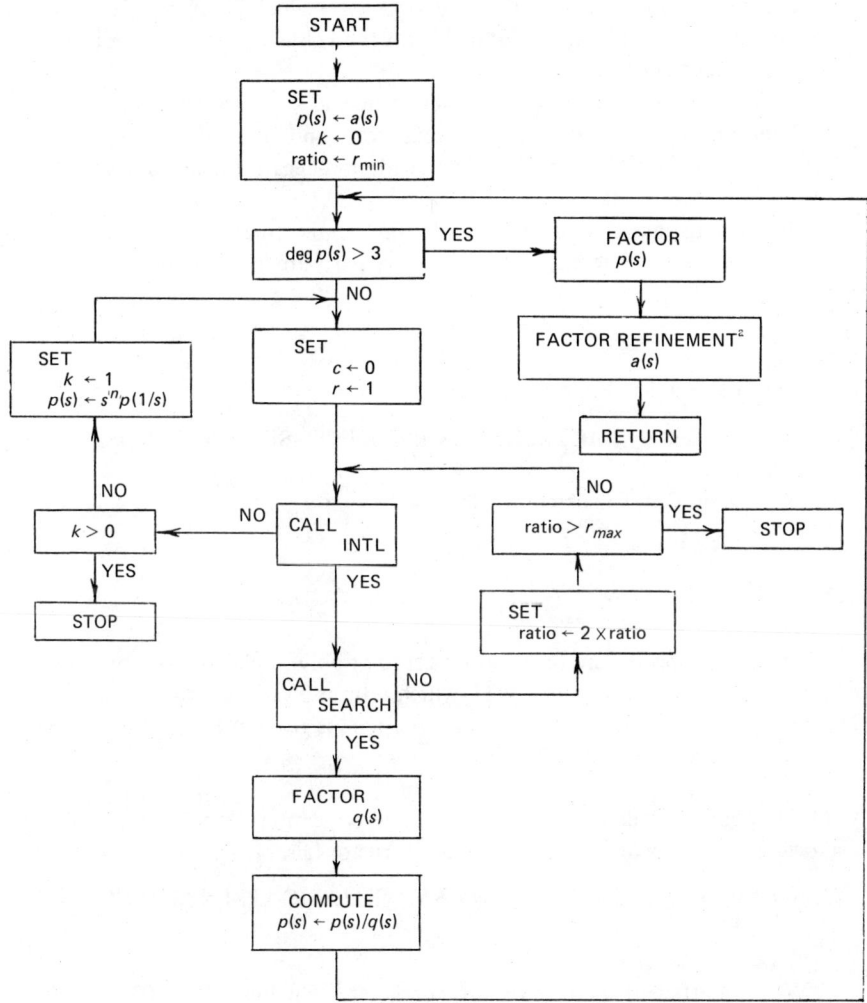

[a]Implemented by Calling Subprogram SEARCH.

Figure 6-2. Computational Flow Chart for Subprogram FACT

using the original polynomial. Subprograms INTL and SEARCH are called for the purposes of initializing and performing the Newton-Raphson search, respectively. Subprogram FACT is summarized as follows:

FACT function: Compute the distinct zero locations for polynomial $a(s)$ as a subprogram.

FACT dummy argument list: N,AV,EPSLN2,PRINT,II,ZM, where
   $1 \leqslant n \leqslant 8$.
   AV(8) = coefficient vector **a** for polynomial $a(s)$ which has N coefficients.
   EPSLN2 = threshold (equals $EPSLN^2$) for the Lehmer-Schur test for containment of zeros.
   PRINT = logical control ("on" for value true and "off" for value false) to determine whether the steps in locating zeros of polynomial $a(s)$ are printed.
   II = number of distinct zero locations, with non-negative imaginary parts, of polynomial $a(s)$ that are found.
   ZM(2,7) = distinct zero locations with non-negative imaginary parts of polynomial $a(s)$ where real and imaginary parts are in rows one and two respectively.

FACT input data list: None.

FACT output data list: None.

Subprograms called by FACT: CONVEX,EVAL,INTL,SEARCH,ZEROS.

   Subprogram FACT is listed in Table 6-2.

INTL, ZEROS, and CONVEX

The initialization procedure for locating a zero of polynomial $p(s)$ is implemented by a logical subprogram INTL which returns a value of true or false. The computational flow chart for this subprogram is given in Figure 6-3. Subprogram INTL is summarized as follows:

INTL Function: Initialize search for distinct zero locations of polynomial $p(s)$ as a logical function which returns a value of true or false.

INTL dummy argument list: N,PV,QV,RV,C,R,RATIO,EPSLN2,PRINT, where
   $1 \leqslant N \leqslant 8$.
   PV(8) = coefficient vector **p** of polynomial $p(s)$ which has N coefficients.
   QV(8) = coefficient vector **q** of polynomial $q(s) = dp(s)/ds$.
   RV(8) = coefficient vector **r** of polynomial $r(s) = dq(s)/ds$.

Table 6-2
Listing of FACT

```
1           SUBROUTINE FACT (N,AV,EPSLN2,PRINT,LL,ZM)
2     C     FACTOR REAL POLYNOMIALS WITH FIRST-ORDER NON-ZERO ZEROS
3           REAL*8 ZM(2,N-1),AV(N),PV(N),QV(N),RV(N)
4           LOGICAL*1 H,INTL,SEARCH,PRINT
5           REAL*8 ZM(2,7),AV(8),PV(8),QV(8),RV(8),D,
6          1E,F,R,RATIO,EPSLN2,EPSLN4,DREAL,DIMAG,DSQRT,DABS
7           COMPLEX*16 C,S,DCONJG,DCMPLX
8     1000  FORMAT (/,4X,'RV')
9     1001  FORMAT (4X,4D18.10)
10    1002  FORMAT (/,4X,'DEFLATION')
11    1003  FORMAT (//,4X,'DEFLATION FAILS')
12    1004  FORMAT (/,4X,'REFINEMENT FAILS')
13    1005  FORMAT (/,4X,'REFINEMENT')
14          EPSLN4=EPSLN2**2
15          L=0
16          LL=0
17          M=N
18          H=.FALSE.
19          DO 1006 I=1,M
20    1006  PV(I)=AV(I)
21    1007  IF (M -3) 1020,1024,1008
22    1008  R=1.D0
23          C=0.D0
24          RATIO=2.D0
25    1009  J=M-1
```

(continued)

Table 6-2 Continued

```
26          DO 1010 I=1,J
27     1010 QV(I)=(M-I)*PV(I)
28          K=J-1
29          DO 1011 I=1,K
30     1011 RV(I)=(J-I)*QV(I)
31          IF (.NOT. PRINT) GO TO 1012
32          WRITE (6,1000)
33          WRITE (6,1001) (PV(I),I=1,M)
34          WRITE (6,1002)
35     1012 IF (INTL(M,PV,QV,RV,C,R,RATIO,EPSLN2,PRINT))
36         1GO TO 1015
37          IF (H) GO TO 1016
38          H=.TRUE.
39          DO 1013 I=1,M
40     1013 QV(I)=PV(I)
41          J=M+1
42          D=PV(M)
43          DO 1014 I=1,M
44     1014 PV(I)=QV(J-I)/D
45          GO TO 1008
46     1015 IF (SEARCH(M,PV,QV,RV,C,S,EPSLN4,PRINT))
47         1GO TO 1017
48          RATIO=2.D0*RATIO
49          IF (RATIO-16.D0) 1012,1012,1016
50     1016 WRITE (6,1003)
51          STOP
```

UTILITY PROGRAMS

```
52   1017  F=DREAL(S*DCONJG(S))
53         D=DIMAG(S)-EPSLN2*DSQRT(F)
54         F=DREAL(S)
55         IF (H)  S=1.D0/S
56         L=L+1
57         LL=LL+1
58         ZM(1,LL)=DREAL(S)
59         IF (D) 1018,1018,1019
60   1018  ZM(2,LL)=0.D0
61         QV(1)=1.D0
62         QV(2)=-E
63         CALL PD (M,PV,2,QV,0.D0,M,PV,KK,QV)
64         GO TO 1007
65   1019  ZM(2,LL)=DABS(DIMAG(S))
66         L=L+1
67         QV(1)=1.D0
68         QV(2)=-2.D0*E
69         QV(3)=F
70         CALL PD (M,PV,3,QV,0.D0,M,PV,KK,QV)
71         GO TO 1007
72   1020  IF (H) GO TO 1021
73         F=PV(2)/PV(1)
74         GO TO 1022
75   1021  F=PV(1)/PV(2)
76   1022  L=L+1
77         LL=LL+1
78         IF (.NOT. PRINT) GO TO 1023
```

(continued)

Table 6-2 Continued

```
79              WRITE (6,1000)
80              WRITE (6,1001) (PV(I),I=1,M)
81       1023   ZM(1,LL)=-F
82              ZM(2,LL)=0.D0
83              GO TO 1030
84       1024   IF (H) GO TO 1025
85              F=PV(2)/PV(1)
86              E=PV(3)/PV(1)
87              GO TO 1026
88       1025   F=PV(2)/PV(3)
89              E=PV(1)/PV(3)
90       1026   D=-.5D0*F
91              E=D*D-E
92              IF (.NOT. PRINT) GO TO 1027
93              WRITE (6,1000)
94              WRITE (6,1001) (PV(I),I=1,M)
95       1027   IF (E) 1028,1029,1029
96       1028   E=DSQRT(-E)
97              L=L+2
98              LL=LL+1
99              ZM(1,LL)=D
100             ZM(2,LL)=E
101             GO TO 1030
102      1029   E=DSQRT(E)
103             L=L+2
104             LL=LL+1
```

UTILITY PROGRAMS

```
105          ZM(1,LL)=D-E
106          ZM(2,LL)=0.D0
107          LL=LL+1
108          ZM(1,LL)=D+E
109          ZM(2,LL)=0.D0
110          IF (L+1-N) 1016,1031,1016
111     1031 IF (N-3) 1036,1036,1032
112     1032 M=N-1
113          DO 1033 I=1,M
114     1033 QV(I)=(N-I)*AV(I)
115          K=M-1
116          DO 1Q34 I=1,K
117     1034 RV(I)=(M-I)*QV(I)
118          H=.FALSE.
119          IF (PRINT) WRITE (6,1005)
120          DO 1035 I=1,LL
121          IF (.NOT. SEARCH(N,AV,QV,RV,DCMPLX(ZM(1,I),
122         1ZM(2,I)),S,EPSLN4,PRINT)) H=.TRUE.
123          ZM(1,I)=DREAL(S)
124     1035 ZM(2,I)=DABS(DIMAG(S))
125          IF (H) WRITE (6,1004)
126     1036 RETURN
127          END
```

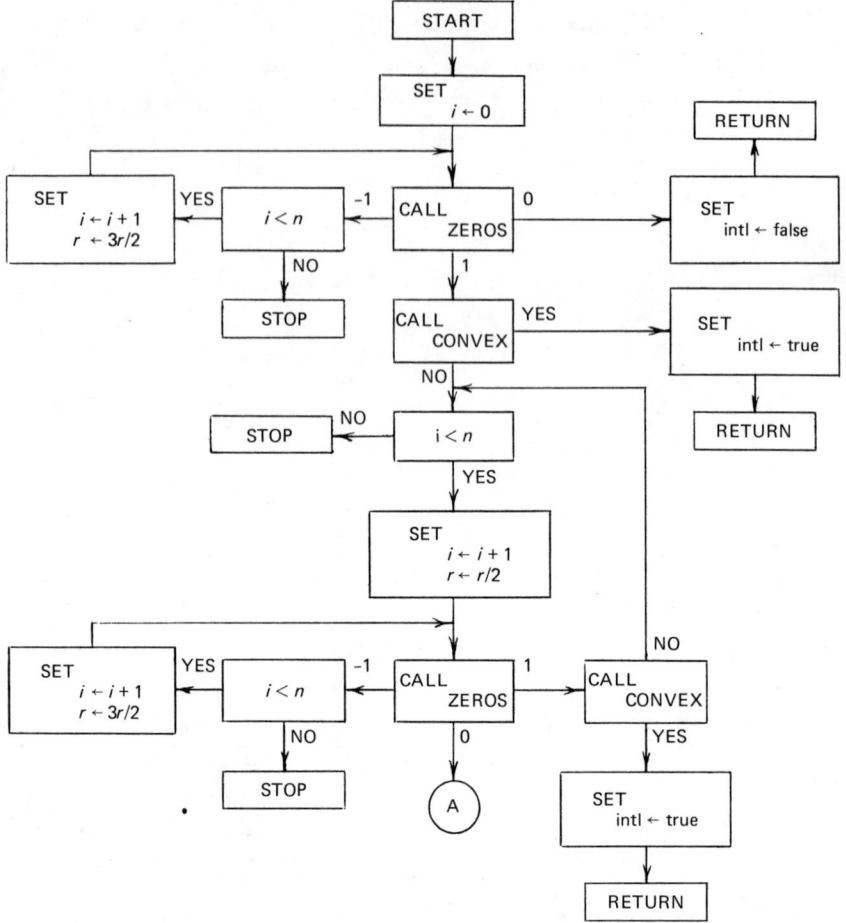

Figure 6-3. Computational Flow Chart for Subprogram INTL

C = complex location of the center of the circle used in the Lehmer-Schur test for containment of zeros.

R = radius of the circle used in the Lehmer-Schur test for containment of zeros.

RATIO = threshold (equals 2, 4, ..., or 16) for ratio $|q(s)|^2/|p(s)r(s)|$ which is used to determine that s lies in a region where function $|p(s)|^2/2$ is convex.

EPSLN2 = threshold (equals $EPSLN^2$) for the Lehmer-Schur test for containment of zeros.

PRINT = logical control ("on" for value true and "off" for value false)

UTILITY PROGRAMS

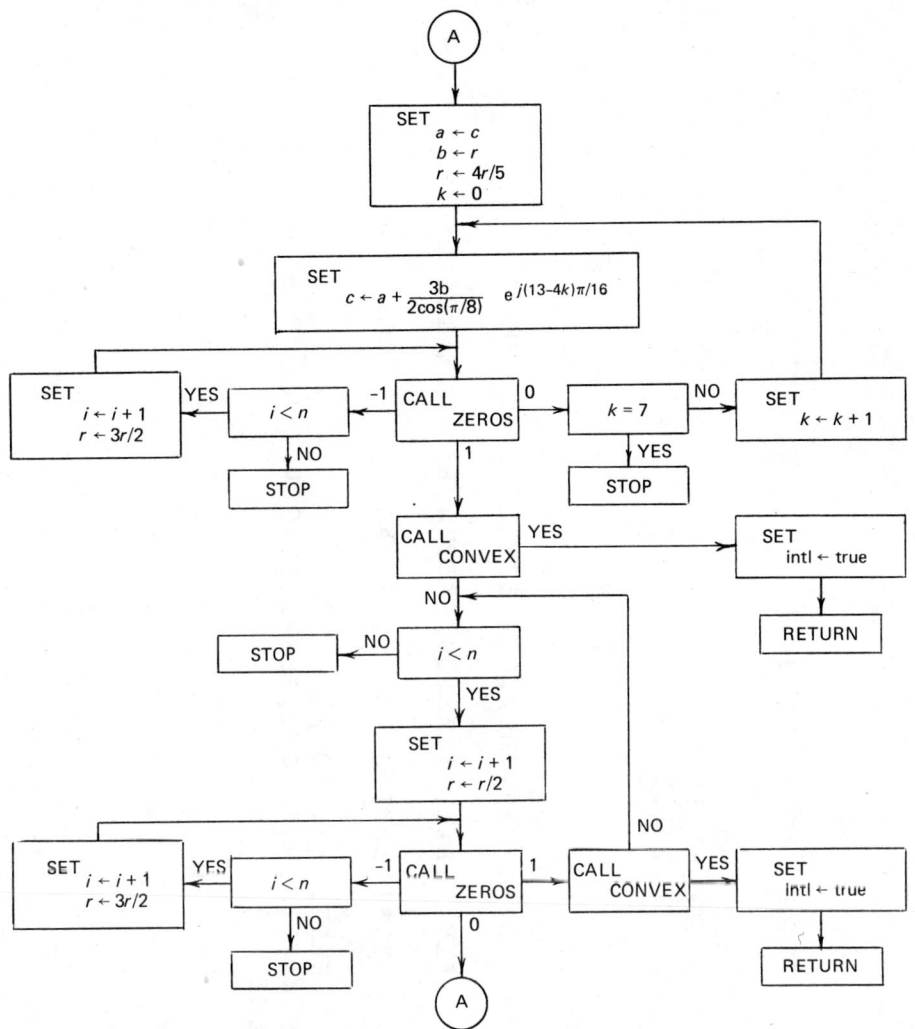

to determine whether the steps in locating zeros of polynomial $p(s)$ are printed.

INTL input data list: None.

INTL output data list: None.

Subprograms called by INTL: CONVEX,EVAL,ZEROS.

Subprogram INTL is also listed in Table 6-3.

Table 6-3
Listing of INTL

```
         LOGICAL FUNCTION INTL*1 (N,PV,QV,RV,C,R,RATIO,
        1                         EPSLN2,PRINT)
C    INITIALIZE SEARCH FOR ZEROS
C
      REAL*8 PV(N),QV(N),RV(N)
      LOGICAL*1 CONVEX,PRINT
      INTEGER*4 ZEROS
      REAL*8 PV(8),QV(8),RV(8),R,EPSLN2,RATIO,A,B,
     1RS,D,E,DCOS,DSIN
      COMPLEX*16 C,CS,DCMPLX
      DATA A/.78539816339744700/,B/1.6235883004385900/
      DATA NI/20/
 1000 FORMAT (4X,'INITIALIZE: ZEROS,RADIUS,REAL(CENTER),',
     1'IMAG(CENTER)')
 1001 FORMAT (//,4X,'LEHMER-SCHUR TEST FAILS')
      I=0
      IF (PRINT) WRITE (6,1000)
 1002 IF (ZEROS(N,PV,C,R,EPSLN2,PRINT)) 1004,1003,1006
 1003 INTL=.FALSE.
      RETURN
 1004 IF (L-NI) 1005,1019,1019
 1005 R=1.5D0*R
      L=L+1
      GO TO 1002
 1006 IF (CONVEX(N,PV,QV,RV,C,RATIO)) GO TO 1026
 1007 IF (L-NI) 1008,1019,1019
```

(continued)

```
26   1008  R=.5D0*R
27   1009  L=L+1
28         IF (ZEROS(N,PV,C,R,EPSLN2,PRINT)) 1015,1010,1025
29   1010  RS=R
30         CS=C
31         R=.8D0*R
32         K=0
33   1011  D=B*RS
34         E=(3.5D0-K)*A
35         C=CS+DCMPLX(D*DCOS(E),D*DSIN(E))
36   1012  IF (ZEROS(N,PV,C,R,EPSLN2,PRINT)) 1017,1013,1020
37   1013  IF (K-7) 1014,1019,1019
38   1014  K=K+1
39         GO TO 1011
40   1015  IF (L-NI) 1016,1019,1019
41   1016  R=1.5D0*R
42         GO TO 1009
43   1017  IF (L-NI) 1018,1019,1019
44   1018  R=1.5D0*R
45         L=L+1
46         GO TO 1012
47   1019  WRITE (6,1001)
48         STOP
49   1020  IF (CONVEX(N,PV,QV,RV,C,RATIO)) GO TO 1026
50   1021  IF (L-NI) 1022,1019,1019
51   1022  R=.5D0*R
52   1023  L=L+1
```

Table 6-3 Continued

```
53          IF (ZEROS(N,PV,C,R,EPSLN2,PRINT)) 1027,1010,1024
54   1024   IF (.NOT. CONVEX(N,PV,QV,RV,C,RATIO))
55         1GO TO 1021
56   1025   IF (.NOT. CONVEX(N,PV,QV,RV,C,RATIO))
57         1GO TO 1007
58   1026   INTL=.TRUE.
59          RETURN
60   1027   IF (L-NI) 1028,1019,1019
61   1028   R=1.5D0*R
62          GO TO 1023
63          END
```

UTILITY PROGRAMS

Subprogram INTL calls another subprogram ZEROS which implements the Lehmer-Schur test for the containment of zeros. Integer subprogram ZEROS returns a value of 1, 0, or -1, and a flow chart for this subprogram is given in Figure 6-4. Subprogram ZEROS is summarized as follows:

ZEROS function: Test for containment of zeros for polynomial $p(s)$ as an integer function which returns a value of 1 (for zeros contained), 0 (for no zeros contained), or -1 (for zeros on the boundary).

ZEROS dummy argument list: N,PV,C,R,EPSLN2,PRINT, where
$1 \leqslant N \leqslant 8$.
PV(8) = coefficient vector **p** of polynomial $p(s)$ which has N coefficients.
C = complex location of the center of the circle used in the Lehmer-Schur test for containment of zeros.
R = radius of the circle used in the Lehmer-Schur test for containment of zeros.
EPSLN2 = threshold (equals $EPSLN^2$) for the Lehmer-Schur test for containment of zeros.
PRINT = logical control ("on" for value true and "off" for value false)

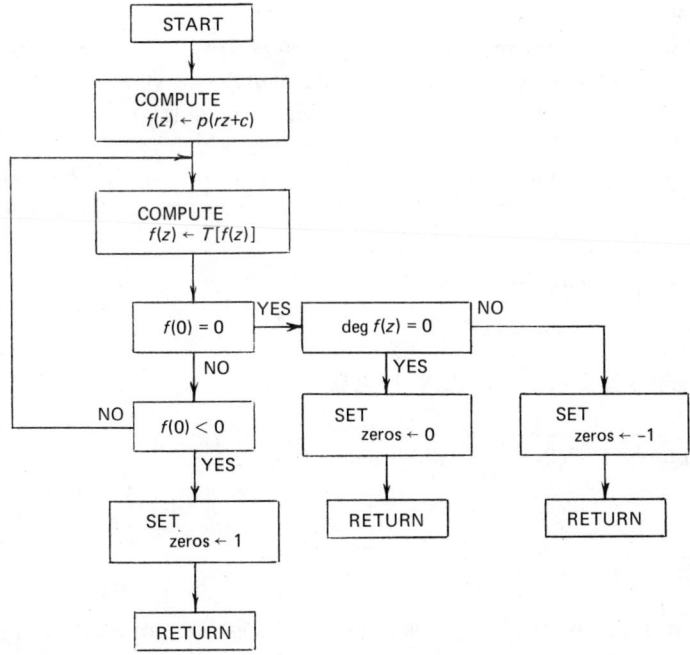

Figure 6-4. Computational Flow Chart for Subprogram ZEROS

to determine whether the steps in locating zeros of polynomial $p(s)$ are printed.

ZEROS input data list: None.

ZEROS output data list: None.

Subprograms called by ZEROS: None.

Subprogram ZEROS is also listed in Table 6-4. The reader should note that subprogram ZEROS calls an IBM 360/65 system-supplied routine ERRSET (208,256,-1,1) which suppresses diagnostic procedures induced by exponent underflow. This routine can be omitted for polynomials of low degree.

Subprogram INTL also calls subprogram CONVEX which implements the test given in equation (6.17) for convexity of function $|p(s)|^2/2$. Logical subprogram CONVEX returns a value of true or false and is summarized as follows:

CONVEX function: Test for convexity of function $|p(s)|^2/2$ as a logical function which returns a value of true or false.

CONVEX dummy argument list: N,PV,QV,RV,S,RATIO, where $1 \leqslant N \leqslant 8$.
PV(8) = coefficient vector **p** of polynomial $p(s)$ which has N coefficients.
QV(8) = coefficient vector **q** of polynomial $q(s) = dp(s)/ds$.
RV(8) = coefficient vector **r** of polynomial $r(s) = dq(s)/ds$.
S = s.
RATIO = threshold (equals 2, 4, ..., or 16) for ratio $|q(s)|^2/|p(s)r(s)|$ which is used to determine that s lies in a region where function $|p(s)|^2/2$ is convex.

CONVEX input data list: None.

CONVEX output data list: None.

Subprograms called by CONVEX: EVAL.

Subprogram CONVEX is also listed in Table 6-5.

SEARCH

The Newton-Raphson search procedure for locating a zero of polynomial $p(s)$ is implemented by a logical subprogram SEARCH which returns a value of true or false. The computational flow chart for this subprogram is given in Figure 6-5.

Table 6-4
Listing of ZEROS

```
 1             INTEGER FUNCTION ZEROS*4  (N,PV,C,R,EPSLN2,PRINT)
 2   C     LEHMER-SCHUR TEST FOR CONTAINMENT OF ZEROS
 3   C     REAL*8 PV(N)
 4   C     COMPLEX*16 FV(N),GV(N)
 5             LOGICAL*1 H,PRINT,HT
 6             REAL*8 PV(8),R,EPSLN2,E,F,DREAL,DIMAG,DMAX1,DABS
 7             COMPLEX*16 FV(8),GV(8),C,D,DCONJG,DCMPLX
 8        1000 FORMAT (4X,I18,3D18.10)
 9             CALL ERRSET (208,256,-1,1)
10             M=N
11             I=M-1
12        1001 FV(1)=PV(1)*R
13             FV(2)=PV(2)+PV(1)*C
14             IF (M-2) 1005,1005,1002
15        1002 DO 1004 I=3,M
16             FV(I)=PV(I)+FV(I-1)*C
17             DO 1003 J=3,I
18             K=I+2-J
19        1003 FV(K)=FV(K)*R+FV(K-1)*C
20        1004 FV(1)=FV(1)*R
21        1005 D=FV(1)
22             DO 1006 I=1,M
23        1006 FV(I)=FV(I)/D
24             L=M+1
25             DO 1007 I=1,M
```

(continued)

Table 6–4 Continued

```
26  1007  GV(I)=DCONJG(FV(L-I))
27        D=DCONJG(FV(M))
28        DO 1008 I=1,M
29  1008  FV(I)=D*FV(I)
30        L=0
31        H=.TRUE.
32        DO 1013 I=1,M
33        HT=.TRUE.
34        D=FV(I)-GV(I)
35        IF (DABS(DREAL(D))-EPSLN2*DMAX1(DABS(DREAL(FV(I))),
36       1DABS(DREAL(GV(I))))) 1009,1009,1010
37  1009  D=DCMPLX(0.D0,DIMAG(D))
38        HT=.FALSE.
39  1010  IF (DABS(DIMAG(D))-EPSLN2*DMAX1(DABS(DIMAG(FV(I))),
40       1DABS(DIMAG(GV(I))))) 1011,1011,1012
41  1011  D=DCMPLX(DREAL(D),0.D0)
42        IF (HT) GO TO 1012
43        IF (H) L=L+1
44        GO TO 1013
45        H=.FALSE.
46  1012  FV(I)=D
47  1013  IF (DREAL(FV(M))) 1020,1014,1014
48  1014  IF (L) 1019,1019,1015
49  1015  M=M-L
50        DO 1016 I=1,M
51  1016  FV(I)=FV(I+L)
```

```
52         GO TO 1005
53    1017 IF (M-1) 1018,1018,1019
54    1018 ZEROS=0
55         GO TO 1021
56    1019 ZEROS=-1
57         GO TO 1021
58    1020 ZEROS=1
59    1021 IF (.NOT. PRINT) RETURN
60         E=DREAL(C)
61         F=DIMAG(C)
62         WRITE (6,1000) ZEROS,R,E,F
63         RETURN
64         END
```

Table 6-5
Listing of CONVEX

```
1         LOGICAL FUNCTION CONVEX*1 (N,PV,QV,RV,S,RATIO)
2    C    TEST CONVEXITY OF POLYNOMIAL FUNCTION
3    C    REAL*8 PV(N),QV(N),RV(N)
4         REAL*8 PV(8),QV(8),RV(8),RATIO,DREAL,
5        1DSQRT,DIMAG
6         COMPLEX*16 D,P,Q,R,S,EVAL,DCONJG
7         P=EVAL(N,PV,S)
8         Q=EVAL(N-1,QV,S)
9         R=EVAL(N-2,RV,S)
10        D=P*DCONJG(R)
11        IF (DREAL(Q*DCONJG(Q))-RATIO*DSQRT(DREAL(D*DCONJG(D))))
12       1 1000,1000,1001
13   1000 CONVEX=.FALSE.
14        RETURN
15   1001 CONVEX=.TRUE.
16        RETURN
17        END
```

UTILITY PROGRAMS

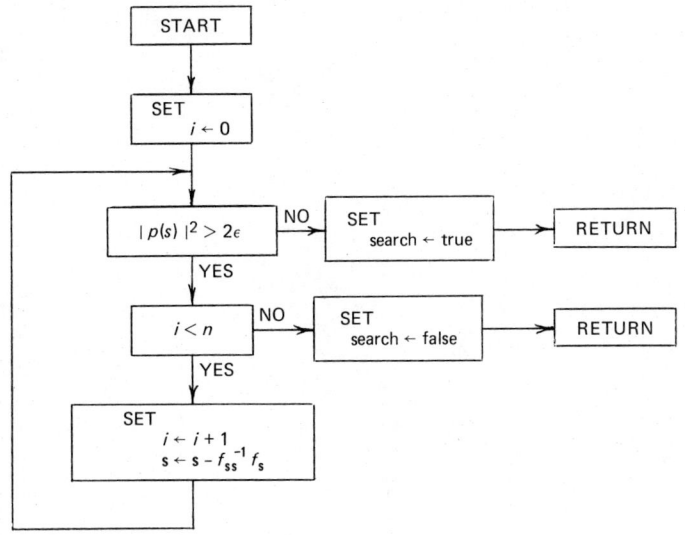

Figure 6-5. Computational Flow Chart for Subprogram SEARCH

Gradient f_s and hessian matrix f_{ss} appearing in Figure 6-5 are given in equations (6.6) and (6.7). Subprogram SEARCH is summarized as follows:

SEARCH function: Newton-Raphson search for a zero location of polynomial $p(s)$ as a logical function which returns a value of true or false.

SEARCH dummy argument list: N,PV,QV,RV,S,SR,EPSLN4,PRINT, where $1 \leqslant N \leqslant 8$.
  PV(8) = coefficient vector **p** of polynomial $p(s)$ which has N coefficients.
  QV(8) = coefficient vector **q** of polynomial $q(s) = dp(s)/ds$.
  RV(8) = coefficient vector **r** of polynomial $r(s) = dq(s)/ds$.
  S = initial value of s.
  SR = returned value of s.
  EPSLN4 = threshold (equals $EPSLN^4$) for terminating iterations in the search for zero locations of the function $|p(s)|^2/2$.
  PRINT = logical control ("on" for value true and "off" for value false) for whether the steps in locating zeros of polynomial $p(s)$ are printed.

SEARCH input data list: None.

SEARCH output data list: None.

Subprograms called by SEARCH: EVAL.

Subprogram SEARCH is also listed in Table 6-6.

Table 6-6
Listing of SEARCH

```
         LOGICAL FUNCTION SEARCH*1 (N,PV,QV,RV,S,SR,
        1                           EPSLN4,PRINT)
C        SEARCH FOR ZEROS
C        REAL*8 PV(N),QV(N),RV(N)
         LOGICAL*1 PRINT
         REAL*8 PV(8),QV(8),RV(8),C,DET,A,B,E,F,
        1EPSLN4,DREAL,DIMAG
         COMPLEX*16 S,P,D,Q,R,EVAL,SR,DCONJG,DCMPLX
         DATA NS/10/
1000     FORMAT (4X,'SEARCH: VALUE,DET; REAL(ROOT),',
        1'IMAG(ROOT)')
1001     FORMAT (4X,4D18.10)
         K=0
         SR=S
         IF (PRINT) WRITE (6,1000)
1002     P=EVAL(N,PV,SR)
         IF (DREAL(P*DCONJG(P))-EPSLN4) 1003,1003,1004
1003     SEARCH=.TRUE.
         RETURN
1004     IF (K-NS) 1006,1005,1005
1005     SEARCH=.FALSE.
         RETURN
1006     Q=EVAL(N-1,QV,SR)
         R=EVAL(N-2,RV,SR)
```

UTILITY PROGRAMS

```
25        D=P*DCONJG(R)
26        C=DREAL(Q*DCONJG(Q))
27        DET=C**2-DREAL(D*DCONJG(D))
28        IF (DET) 1005,1005,1007
29  1007  A=DREAL(D)
30        B=DIMAG(D)
31        D=P*DCONJG(Q)
32        E=DREAL(D)
33        F=DIMAG(D)
34        SR=SR-DCMPLX(((C-A)*E-B*F)/DET,
35       1(-B*E+(A+C)*F)/DET)
36        K=K+1
37        IF (.NOT. PRINT) GO TO 1002
38        C=.5D0*DREAL(P*DCONJG(P))
39        E=DREAL(SR)
40        F=DIMAG(SR)
41        WRITE (6,1001) C,DET,E,F
42        GO TO 1002
43        END
```

EVAL, MATCH, PC, PD, and PM

Subprograms FACTOR and FACT call a number of relatively simple utility subprograms which are summarized briefly here.

Subprogram EVAL evaluates the numerical value of polynomial $p(s)$ for a given value of s using Horner's method [11]. This floating-point subprogram returns a complex value and is summarized as follows:

EVAL function: Compute the value of polynomial $p(s)$ as a complex floating-point function which returns a complex value.

EVAL dummy argument list: N,PV,S, where
$1 \leqslant N \leqslant 8$.
PV(8) = coefficient vector **p** of polynomial $p(s)$ which has N coefficients.
S = s.

EVAL input data list: None.

EVAL output data list: None.

Subprograms called by EVAL: None.

Subprogram EVAL is also listed in Table 6-7.

Table 6-7
Listing of EVAL

```
1              COMPLEX FUNCTION EVAL*16 (N,PV,S)
2      C       EVALUATE REAL POLYNOMIAL WITH A COMPLEX ARGUMEN'
3      C       REAL*8 PV(N)
4              REAL*8 PV(8)
5              COMPLEX*16 S
6              IF (N-1) 1000,1001,1002
7       1000   EVAL=0.D0
8              RETURN
9       1001   EVAL=PV(1)
10             RETURN
11      1002   EVAL=PV(1)
12             J=N-1
13             DO 1003 I=1,J
14      1003   EVAL=S*EVAL+PV(I+1)
15             RETURN
16             END
```

UTILITY PROGRAMS

Subprogram MATCH matches zero locations between polynomials $p(s)/q(s)$ and $q(s)/r(s)$ where

$$q(s) = p(s) \text{ gcd } dp(s)/ds \quad \text{and} \quad r(s) = q(s) \text{ gcd } dq(s)/ds.$$

This logical subprogram returns a value of true or false and is summarized as follows:

MATCH function: Match two zero locations as a logical function which returns true or false.

MATCH dummy argument list: M,N,ZM,D,E,L, where
$1 \leqslant M \leqslant N \leqslant 7$.
ZM(2,7) = distinct zero locations with non-negative imaginary parts where real and imaginary parts are in rows one and two respectively.
D = real part of a zero location.
E = non-negative imaginary part of a zero location.
L = column number of ZM for matching a zero location.

MATCH input data list: None.

MATCH output data list: None.

Subprograms called by MATCH: None.

Subprogram MATCH is also listed in Table 6-8.

Subprogram PC computes the greatest polynomial common divisor $c(s) = a(s) \text{ gcd } b(s)$. A computational flow chart for this subprogram is given in Figure 6.6 and is summarized as follows:

PC function: Compute the greatest common polynomial divisor $c(s)$ of $a(s)$ and $b(s)$ as a subprogram.

PC dummy argument list: NA,AV,NB,BV,EPSLN,NC,CV, where
$1 \leqslant NA \leqslant 8$.
AV(8) = coefficient vector **a** of polynomial $a(s)$ which has NA coefficients.
$1 \leqslant NB \leqslant 8$.
BV(8) = coefficient vector **b** of polynomial $b(s)$ which has NB coefficients.
EPSLN = threshold for computing the degree of $c(s)$.
$1 \leqslant NC \leqslant 8$.
CV(8) = coefficient vector **c** of polynomial $c(s)$ which has NC coefficients.

PC input data list: None.

PC output data list: None.

Subprograms called by PC: PD.

Table 6-8
Listing of MATCH

```
1              LOGICAL FUNCTION MATCH*1 (M,N,ZM,D,E,L)
2       C      MATCH TWO ZERO LOCATIONS
3       C      REAL*8 ZM(2,N-1)
4              REAL*8 ZM(2,7),D,E,DABS,DMAX1,F,G
5              F=1.D75
6              DO 1001 I=M,N
7              G=DABS(D-ZM(1,I))+DABS(E-ZM(2,I))
8              IF (G-F) 1000,1001,1001
9       1000   F=G
10             L=I
11      1001   CONTINUE
12             IF (D) 1004,1002,1004
13      1002   IF (ZM(1,L)) 1003,1004,1003
14      1003   MATCH=.FALSE.
15             RETURN
16      1004   IF (E) 1007,1005,1007
17      1005   IF (ZM(2,L)) 1006,1007,1006
18      1006   MATCH=.FALSE.
19             RETURN
20      1007   MATCH=.TRUE.
21             RETURN
22             END
```

Subprogram PC is also listed in Table 6-9.

Subprogram PD computes the polynomial quotient $c(s)$ and remainder $d(s)$ which appear in

$$c(s) + \frac{d(s)}{b(s)} = \frac{a(s)}{b(s)}$$

for given polynomials $a(s)$ and $b(s)$. This subprogram is summarized as follows:

PD function: Compute the quotient and remainder polynomials corresponding to

$$c(s) + \frac{d(s)}{b(s)} = \frac{a(s)}{b(s)}$$

as a subprogram.

UTILITY PROGRAMS

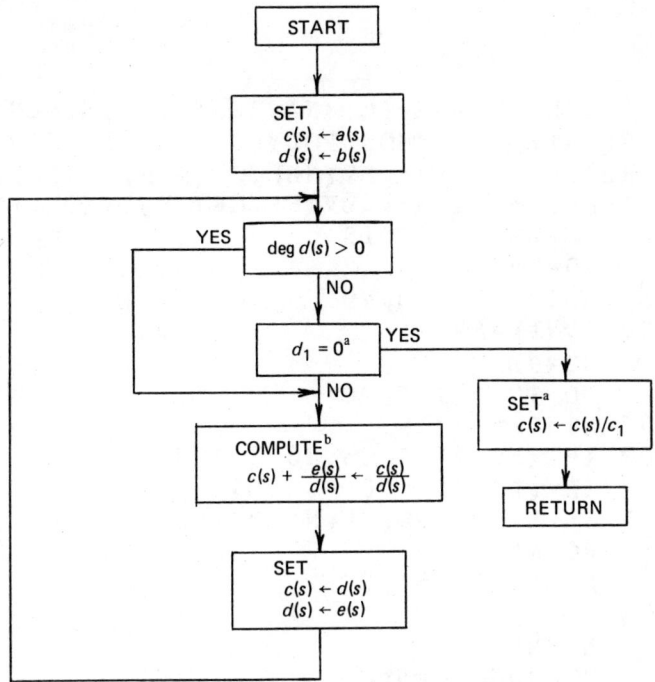

[a] c_1 and d_1 are leading coefficients of $c(s)$ and $d(s)$ respectively.
[b] Implemented by calling subprogram PD.

Figure 6-6. Computational Flow Chart for Subprogram PC

PD dummy argument list: NA,AV,NB,EPSLN,BV,NC,CV,ND,DV, where
$1 \leqslant NA \leqslant 8$.
AV(8) = coefficient vector **a** of polynomial $a(s)$ which has NA coefficients.
$1 \leqslant NB \leqslant 8$.
BV(8) = coefficient vector **b** of polynomial $b(s)$ which has NB coefficients.
EPSLN = threshold used to computing the degree of $d(s)$.
$1 \leqslant NC \leqslant 8$.
CV(8) = coefficient vector **c** of polynomial $c(s)$ which has NC coefficients.
$1 \leqslant ND \leqslant 8$.
DV(8) = coefficient vector **d** of polynomial $d(s)$ which has ND coefficients.

PD input data list: None.

PD output data list: None.

Subprograms called by PD: None.

This subprogram is also listed in Table 6-10.

Table 6-9
Listing of PC

```
 1           SUBROUTINE PC (NA,AV,NB,BV,EPSLN,NC,CV)
 2     C     GREATEST COMMON POLYNOMIAL DIVISOR
 3     C     REAL*8 AV(NA),BV(NB),CV(NC),DV(ND),EV(NB)
 4           REAL*8 AV(8),BV(8),CV(8),DV(8),EV(8)
 5           REAL*8 D,EPSLN
 6           ND=NA
 7           DO 1000 I=1,ND
 8     1000  DV(I)=AV(I)
 9           NC=NB
10           DO 1001 I=1,NC
11     1001  CV(I)=BV(I)
12     1002  IF (ND-1) 1003,1003,1004
13     1003  IF (DV(1)) 1004,1007,1004
14     1004  CALL PD (NC,CV,ND,DV,EPSLN,NC,CV,NE,EV)
15           NC=ND
16           DO 1005 I=1,NC
17     1005  CV(I)=DV(I)
18           ND=NE
19           DO 1006 I=1,ND
20     1006  DV(I)=EV(I)
21           GO TO 1002
22     1007  D=CV(1)
23           DO 1008 I=1,NC
24     1008  CV(I)=CV(I)/D
25           RETURN
26           END
```

Subprogram PM computes the polynomial product $c(s) = a(s)b(s)$ and is summarized as follows:

PM function: Compute the polynomial product

$$c(s) = a(s)b(s)$$

as a subprogram.

PM dummy argument list: NA,AV,NB,BV,NC,CV where
 $1 \leqslant NA \leqslant 8$.
 AV(8) = coefficient vector a of polynomial $a(s)$ which has NA coefficients.
 $1 \leqslant NB \leqslant 8$.

```
      SUBROUTINE PD (NA,AV,NB,BV,EPSLN,NC,CV,ND,DV)
C     POLYNOMIAL QUOTIENT AND REMAINDER
C     REAL*8 AV(NA),BV(NB),CV(NC),DV(ND),EV(NE)
      REAL*8 AV(8),BV(8),CV(8),DV(8),EV(8)
      REAL*8 D,E,F,EPSLN,DABS,DMAX1
      NE=NB
      DO 1000 I=1,NE
 1000 EV(I)=BV(I)/BV(1)
      ND=NA
      NC=NA-NB+1
      DO 1001 I=1,ND
 1001 DV(I)=AV(I)
      IF (NC) 1012,1012,1002
 1002 DO 1005 I=1,NC
      D=DV(I)
      K=I-1
      DO 1004 J=1,NE
      L=K+J
      F=D*EV(J)
      E=DV(L)-F
      IF (DABS(E)-EPSLN*DMAX1(DABS(DV(L)),DABS(F)))
     1 1003,1003,1004
 1003 E=0.D0
 1004 DV(L)=E
 1005 CV(I)=D
      J=NC+1
 1006 IF (J-NA) 1007,1007,1011
 1007 IF (DV(J)) 1009,1008,1009
 1008 J=J+1
      GO TO 1006
 1009 J=J-1
      ND=NA-J
      DO 1010 I=1,ND
 1010 DV(I)=DV(I+J)
      RETURN
 1011 ND=1
      DV(1)=0.D0
      RETURN
 1012 ND=NA
      NC=1
      CV(1)=0.D0
      RETURN
      END
```

Table 6-11
Listing of PM

```
 1          SUBROUTINE PM  (NA,AV,NB,BV,NC,CV)
 2    C     POLYNOMIAL PRODUCT
 3    C     REAL*8 AV(NA),BV(NB),CV(NC),DV(NC)
 4          REAL*8 AV(8),BV(8),CV(8),DV(8)
 5          ND=NA+NB-1
 6          DO 1000 I=1,ND
 7     1000 DV(I)=0.
 8          DO 1001 I=1,NA
 9          K=I-1
10          DO 1001 J=1,NB
11          L=K+J
12     1001 DV(L)=DV(L)+AV(I)*BV(J)
13          NC=ND
14          DO 1002 I=1,NC
15     1002 CV(I)=DV(I)
16          RETURN
17          END
```

BV(8) = coefficient vector **b** of polynomial $b(s)$ which has NB coefficients. $1 \leq NC \leq 8$.

CV(8) = coefficient vector **c** of polynomial $c(s)$ which has NC coefficients.

PM input data list: None.

PM output data list: None.

Subprograms called by PM: None.

Subprograms PM is also listed in Table 6-11.

TEST

Program TEST is included as a system-supplied program for the purpose of permitting the user to experiment with subprogram FACTOR in a convenient fashion. Polynomials to be factored can be formed easily with multiple factors, and initialization and search steps can be printed out in order to observe application of the Lehmer-Schur test and to observe quadratic convergence of the Newton-Raphson method. Program TEST is summarized as follows:

TEST function: Test polynomial factoring program FACTOR.

UTILITY PROGRAMS 321

TEST dummy argument list: None.

TEST input data list:

$$\text{C1/C2/C3/GAIN,RADIUS,EPSLN/TRACE}/\underbrace{N/NQ/QV}_{\text{N times}}$$
$$\text{indefinite number of times}$$

where
  GAIN = order of magnitude of the leading coefficient which is used to determine the degree of the polynomial to be factored after zeros at the origin have been removed.
  RADIUS = order of magnitude of the geometric mean of nonzero radii of zero locations which is used to normalize the polynomial to have a unit geometric mean of zero locations.
  EPSLN = threshold for treating zero locations as distinct.
  TRACE = logical control ("on" for value true and "off" for value false) to determine whether the steps in locating zeros are traced by printing.
  N = number of component polynomials.
  $1 \leqslant NQ \leqslant 8$.
  QV(8) = coefficient vector of component polynomial $q(s)$ which has NQ coefficients.

TEST output data list: None.

Subprograms called by TEST: CONVEX,EVAL,FACT,FACTOR,INTL,MATCH, PC,PD,PM,SEARCH,ZEROS.

  Program TEST is also listed in Table 6-12.

Example 9

An example of the use of program TEST is provided by polynomial

$$p(s) = (s+1)(s+2)(s+3)(s+4).$$

Input data for this example is listed in Table 6-13a which specifies that each factor of $p(s)$ is entered as a polynomial $q(s)$.
  Corresponding program output is listed in Table 6-13b. Heading RV denotes coefficient vector **r** of the polynomial $r(s)$ which remains after polynomial $p(s)$ is deflated by factors corresponding to zeros located already found. Also, heading VALUE denotes the numerical value of function $p(s)p(s^*)/2$.

Table 6-12
Listing of TEST

```
 1    C      TEST
 2    C      INTEGER*4 MV(N-1)
 3    C      REAL*8 ZM(2,N-1),AV(N),BV(N)
 4           INTEGER*4 MV(7)
 5           LOGICAL*1 PRINT,COM(65)
 6           REAL*8 ZM(2,7),AV(8),BV(8),GAIN,RADIUS,EPSLN
 7    1000 FORMAT ('1')
 8    1001 FORMAT (65A1)
 9    1002 FORMAT (4X,65A1)
10    1003 FORMAT (G10)
11    1004 FORMAT (4D18.10)
12    1005 FORMAT (//,4X,'TRACE')
13    1006 FORMAT (4X,'GAIN,RADIUS,EPSLN')
14    1007 FORMAT (4X,4D18.10)
15    1008 FORMAT (I10)
16    1009 FORMAT (//,4X,'N')
17    1010 FORMAT (4X,'NQ')
18    1011 FORMAT (4X,'QV')
19    1012 FORMAT (/,4X,'PV')
20    1013 FORMAT (/,4X,'REAL PART,IMAGINARY PART
21         1         'MULTIPLICITY')
22    1014 FORMAT (4X,2D18.10,I18)
23           WRITE (6,1000)
24           DO 1015 I=1,3
25           READ (5,1001) COM
26    1015 WRITE (6,1002) COM
27           READ (5,1003) PRINT
28           READ (5,1004) GAIN,RADIUS,EPSLN
29           WRITE (6,1005)
30           WRITE (6,1003) PRINT
31           WRITE (6,1006)
32           WRITE (6,1007) GAIN,RADIUS,EPSLN
33    1016 READ (5,1008,END=1019) K
34           WRITE (6,1009)
35           WRITE (6,1008) K
36           N=1
37           AV(1)=1.D0
38           DO 1017 I=1,K
39           READ (5,1008) M
40           WRITE (6,1010)
```

Table 6-12 Continued

```
41              WRITE (6,1008) M
42              READ (5,1004) (BV(J),J=1,M)
43              WRITE (6,1011)
44              WRITE (6,1007) (BV(J),J=1,M)
45       1017   CALL PM (N,AV,M,BV,N,AV)
46              WRITE (6,1012)
47              WRITE (6,1007) (AV(I),I=1,N)
48              CALL FACTOR (N,AV,GAIN,RADIUS,EPSLN,PRINT,
49       1                   LL,ZM,MV)
50              WRITE (6,1013)
51              DO 1018 I=1,LL
52       1018   WRITE (6,1014) ZM(1,I),ZM(2,I),MV(I)
53              GO TO 1016
54       1019   WRITE (6,1000)
55              STOP
56              END
```

Finally, heading DET denotes the determinant of the hessian matrix appearing in equation (6.7) which can be written as

$$|f_{ss}| = |q(s)q(s^*)|^2 - |p(s)r(s^*)|^2. \qquad (6.19)$$

These data clearly demonstrate quadratic convergence of the Newton-Raphson method of successive substitutions as well as the overall accuracy of this method of factoring polynomials.

6.2 Solving Linear Algebraic Equations

The computational methods which are implemented by computer programs described in this section also fall into two general categories. The first category involves methods for solving linear algebraic equations in various modified forms of matrix Lyapunov equations. The second category involves matrix inversion.

Description of the Basic Methods

The first two computer programs described here concern the solution of

$$XA + BX + C = 0. \qquad (6.20)$$

Table 6-13a
Listing of User-supplied Input Data for Example 9

```
 1   TEST (CONVEX,EVAL,FACT,FACTOR,INTL,MATCH,PC,PD,PM,SEARCH,ZEROS)
 2   EXAMPLE 9   MERRIAM
 3   UNIVERSITY OF ROCHESTER
 4              T                                      3.    D-08
 5   1.         1.
 6        4
 7        2
 8   1.        1.
 9        2
10   1.        2.
11        2
12   1.        3.
13        2
14   1.        4.
```

Table 6-13b
Printed Output from TEST and Subprograms for Example 9

```
TEST (CONVEX,EVAL,FACT,FACTOR,INTL,MATCH,PC,PD,PM,SEARCH,ZEROS)
EXAMPLE 9   MERRIAM
UNIVERSITY OF ROCHESTER


TRACE     T
GAIN,RADIUS,EPSLN
 0.1000000000D 01    0.1000000000D 01    0.3000000000D-07

N      4

NQ     2

QV
 0.1000000000D 01    0.1000000000D 01

NQ     2

QV
 0.1000000000D 01    0.2000000000D 01

NQ     2

QV
 0.1000000000D 01    0.3000000000D 01
```

(continued)

Table 6–13 Continued

NQ
 2

QV
 0.1000000000D 01   0.4000000000D 01

PV
 0.1000000000D 01   0.1000000000D 02   0.3500000000D 02   0.5000000000D 02
 0.2400000000D 02

REMAINDER POLY FOR MULTIPLE ZERO DEFLATION
 0.0000000000D 00

RV
 0.1000000000D 01   0.1000000000D 02   0.3500000000D 02   0.5000000000D 02
 0.2400000000D 02

DEFLATION
INITIALIZE: ZEROS, RADIUS, REAL(CENTER), IMAG(CENTER)

-1	0.1000000000D 01	0.0000000000D 00	0.0000000000D 00	0.0000000000D 00
1	0.1500000000D 01	0.0000000000D 00	0.0000000000D 00	0.0000000000D 00
0	0.7500000000D 00	0.0000000000D 00	0.0000000000D 00	0.0000000000D 00
1	0.6000000000D 00	-0.1125000000D 01	0.4659902577D 00	0.4659902577D 00
0	0.3000000000D 00	-0.1125000000D 01	0.4659902577D 00	0.4659902577D 00
0	0.2400000000D 00	-0.1575000000D 01	0.6523863607D 00	0.6523863607D 00
0	0.2400000000D 00	-0.1311396103D 01	0.9159902577D 00	0.9159902577D 00
0	0.2400000000D 00	-0.9386038969D 00	0.9159902577D 00	0.9159902577D 00

```
            0             0.2400000000D 00   -0.6750000000D 00    0.6523863607D 00
            0             0.2400000000D 00   -0.6750000000D 00    0.2795941546D 00
            1             0.2400000000D 00   -0.9386038969D 00    0.1599025767D-01
SEARCH: VALUE,DET;        REAL(ROOT),IMAG(ROOT)
 0.9032493910D-01         0.2932797871D 04   -0.9852627068D 00    0.2739556038D-02
 0.4267827651D-02         0.1599466152D 04   -0.9989066393D 00    0.1371193522D-03
 0.2194407170D-04         0.1316905780D 04   -0.9999934417D 00    0.5464258032D-06
 0.7796013074D-09         0.1296124665D 04   -0.9999999998D 00    0.1313952183D-10
 0.1014996447D-17         0.1296000005D 04   -0.1000000000D 01    0.1142301105D-19
RV
 0.1000000000D 01         0.9000000000D 01    0.2600000000D 02    0.2400000000D 02

DEFLATION
INITIALIZE: ZEROS,RADIUS,REAL(CENTER),IMAG(CENTER)
            0             0.1000000000D 01    0.0000000000D 00    0.0000000000D 00

RV
 0.1000000000D 01         0.1083333333D 01    0.3750000000D 00    0.4166666667D-01

DEFLATION
INITIALIZE: ZEROS,RADIUS,REAL(CENTER),IMAG(CENTER)
            1             0.1000000000D 01    0.0000000000D 00    0.0000000000D 00
            1             0.5000000000D 00    0.0000000000D 00    0.0000000000D 00
           -1             0.2500000000D 00    0.0000000000D 00    0.0000000000D 00
            1             0.3750000000D 00    0.0000000000D 00    0.0000000000D 00
            0             0.1875000000D 00    0.0000000000D 00    0.0000000000D 00
```

(continued)

Table 6-13 Continued

```
                    1    0.150000000D 00  -0.281250000C0D 00   0.116497564 4D 00
                    0    0.750000000D-01  -0.281250000 00D 00   0.116497564 4D 00
                    0    0.600000000D-01  -0.393750000 00D 00   0.163096590 2D 00
                    0    0.600000000D-01  -0.327849025 8D 00   0.228997564 4D 00
                    0    0.600000000D-01  -0.234650974 2D 00   0.228997564 4D 00
                    0    0.600000000D-01  -0.168750000 00D 00   0.163096590 2D 00
                    0    0.600000000D-01  -0.168750000 00D 00   0.698985386 5D-01
                    1    0.600000000D-01  -0.234650974 2D 00   0.399756441 7D-02

SEARCH: VALUE,DET:    REAL(ROOT),IMAG(ROOT)
0.864089894 8D-07    0.931984794 1D-06  -0.244279607D 00   0.120265500 7D-02
0.886197442 5D-08    0.366833312 6D-06  -0.248846482 5D 00   0.175408211 4D-03
0.306482178 6D-09    0.217664981 7D-06  -0.249940243 7D 00   0.614077301 4D-05
0.784599500 0D-12    0.189824836 1D-06  -0.249999828 7D 00   0.117088777 8D-07
0.639920573 9D-17    0.188384242 9D-06  -0.250000000 0D 00   0.641900641 7D-13
0.433004262 2D-27    0.188380111 9D-06  -0.250000000 0D 00   0.289855264 5D-23

RV  0.100000000 0D 01    0.833333333 3D 00    0.166666666 7D 00

REFINEMENT
SEARCH: VALUE,DET:    REAL(ROOT),IMAG(ROOT)
0.305446942 5D-26    0.129600000 0D 04  -0.400000000 0D 01  -0.000000000 0D 00
0.908767762 8D-27    0.129600000 0D 04  -0.400000000 0D 01  -0.000000000 0D 00
SEARCH: VALUE,DET:    REAL(ROOT),IMAG(ROOT)
0.141994962 9D-26    0.160000000 0D 02  -0.300000000 0D 01  -0.000000000 0D 00
```

UTILITY PROGRAMS

```
SEARCH: VALUE,DET;    REAL(ROOT),IMAG(ROOT)
 0.4038967835D-27      0.1600000000D 02  -0.2000000000D 01  -0.0000000000D 00
 0.1009741959D-27      0.1600000000D 02  -0.2000000000D 01  -0.0000000000D 00
 0.2524354897D-28      0.1600000000D 02  -0.2000000000D 01  -0.0000000000D 00

REAL PART, IMAGINARY PART, MULTIPLICITY
-0.1000000000D 01      0.0000000000D 00   1
-0.4000000000D 01      0.0000000000D 00   1
-0.3000000000D 01      0.0000000000D 00   1
-0.2000000000D 01      0.0000000000D 00   1
```

Subprogram LIN1 implements Theorem 4-2 of the book *Automated Design of Control Systems* [1]. Although the computational technique implemented in subprogram LIN1 is fast, significant numerical errors may be experienced for matrices of large dimensions. These errors arise in the computation of coefficients of the characteristic polynomial using the method given in equations (3.13) and (3.14). A somewhat slower but more accurate technique [12] for solving equation (6.20) is implemented in subprogram LIN1A. Except for possible numerical errors, subprograms LIN1 and LIN1A can be used interchangeably.

The next two subprograms described here concern the solution of

$$XA + A'X + \sum_{k=1}^{m} B_k' X B_k + C = 0$$

by first reducing these equations to standard vector-matrix form

$$\Delta x + c = 0. \qquad (6.22)$$

Reduction is accomplished by defining vectors **x** and **c** in partitioned form by

$$X = [x_1 \; x_2 \ldots x_n] \text{ and } C = [c_1 \; c_2 \ldots c_n] \qquad (6.23)$$

and then computing the equivalent coefficient matrix Δ. The final solution is obtained by inverting matrix Δ and computing

$$x = \Delta^{-1} c. \qquad (6.24)$$

The last two subprograms described here implement matrix inversion. Subprogram MI implements the pivot method of matrix inversion which can be summarized as follows: Suppose equation

$$Ax = y \qquad (6.25)$$

is written in partitioned form

$$\begin{bmatrix} A_1 & a_1 & A_2 \\ a_2' & a_{mn} & a_3' \\ A_3 & a_4 & A_4 \end{bmatrix} \begin{bmatrix} x_1 \\ x_n \\ x_2 \end{bmatrix} = \begin{bmatrix} y_1 \\ y_m \\ y_2 \end{bmatrix}. \qquad (6.26)$$

Then variable x_n can be eliminated by solving the mth equation appearing in (6.26), assuming coefficient a_{mn} is not zero, to obtain

UTILITY PROGRAMS

$$\begin{bmatrix} \left(A_1 - \dfrac{a_1 a_2'}{a_{mn}}\right) & \dfrac{a_1}{a_{mn}} & \left(A_2 - \dfrac{a_1 a_3'}{a_{mn}}\right) \\ -\dfrac{a_2'}{a_{mn}} & \dfrac{1}{a_{mn}} & -\dfrac{a_3'}{a_{mn}} \\ \left(A_3 - \dfrac{a_4 a_2'}{a_{mn}}\right) & \dfrac{a_4}{a_{mn}} & \left(A_4 - \dfrac{a_4 a_3'}{a_{mn}}\right) \end{bmatrix} \begin{bmatrix} x_1 \\ y_m \\ x_2 \end{bmatrix} = \begin{bmatrix} y_1 \\ x_n \\ y_2 \end{bmatrix} \quad (6.27)$$

This process can be repeated using a different equation each time until all of the variables are eliminated. After a suitable reordering of the rows and columns of the resulting coefficient matrix, matrix A^{-1} is obtained. Subprogram MIA implements the square root decomposition method for inverting a symmetric positive definite matrix [7]. Such a matrix can be put into the form

$$A = B'B \quad (6.28)$$

where matrix

$$B = \begin{bmatrix} b_{11} & b_{12} & \cdots & b_{1n} \\ 0 & b_{22} & \cdots & b_{2n} \\ \cdots & \cdots & \cdots & \cdots \\ 0 & 0 & \cdots & b_{nn} \end{bmatrix} \quad (6.29)$$

is upper triangular and easily computed. An upper triangular matrix is easily inverted computationally so that the desired result is obtained in the form

$$A^{-1} = (B^{-1})(B^{-1})'. \quad (6.30)$$

Subprograms MI and MIA are interchangeable for symmetric positive definite matrices, which is in the most frequent case of matrix inversion encountered in system-supplied programs.

LIN1 and LIN1A

Subprograms LIN1 and LIN1A are used to solve (6.20) and are summarized as follows:

LIN1 and LIN1A function: Compute the solution to

$$XA + BX + C = 0$$

as a subprogram.

LIN1 and LIN1A dummy argument list: MT,NT,M,N,AM,BM,CM,XM, where
$1 \leqslant MT \leqslant 7, 1 \leqslant NT \leqslant 7, 1 \leqslant M \leqslant MT, 1 \leqslant N \leqslant NT$.
AM(NT,NT) = **A**.
BM(MT,MT) = **B**.
CM(MT,NT) = **C**.
XM(MT,NT) = **X**.

LIN1 and LIN1A input data list: None.

LIN1 and LIN1A output data list: None.

Subprograms called by LIN1 and LIN1A: MI.

Subprogram LIN1 is listed in Table 6-14; subprogram LIN1A, in Table 6-15.

LIN2 and INC

Subprogram LIN2 implements equations (6.23) and (6.24) in the solution of (6.21), once matrix Δ^{-1} is returned from a call to subprogram INC via the argument list of LIN2. Subprogram LIN2 is summarized as follows:

LIN2 Function: Compute the solution to

$$XA + A'X + \sum_{k=1}^{m} B_k' X B_k + C = 0$$

as a subprogram.

LIN2 dummy argument list: MT,MX,NX,ZM,CM,XM, where
$1 \leqslant MT \leqslant 28, 1 \leqslant MX \leqslant 7, 1 \leqslant NX \leqslant MX$.
ZM(MT,MT) = reduced equivalent coefficient matrix.
CM(MX,MX) = **C**.
XM(MX,MX) = **X**.

LIN2 input data list: None.

LIN2 output data list: None.

Subprograms called by LIN2: None.

Subprogram LIN2 also is listed in Table 6-16.

Table 6-14
Listing of LIN1

```
         SUBROUTINE LIN1(MT,NT,M,N,AM,BM,CM,XM)
   C     SOLUTION OF XM*AM+BM*XM+CM=0 BY SPECTRAL FACTORIZATION
   C     REAL*8 AM(N,N),BM(M,M),CM(M,N),DM(MN,MN),LM(N,N),
   C    1MM(M,M),NM(M,M),XM(M,N),D,E
         REAL*8 AM(NT,NT),BM(MT,MT),CM(MT,NT),XM(MT,NT),D,E
         REAL*8 DM(7,7),MM(7,7),NM(7,7),LM(7,7)
         DATA MN/7/
         E=1.D0
         DO 1001 I=1,M
         DO 1000 J=1,M
         NM(I,J)=0.D0
1000     MM(I,J)=0.D0
         MM(I,I)=1.D0
         DO 1001 J=1,N
1001     XM(I,J)=0.D0
         DO 1002 I=1,N
         DO 1002 J=1,N
1002     LM(I,J)=0.D0
         I=0
1003     L=L+1
         DO 1006 I=1,N
         DO 1005 J=1,N
         D=0.D0
         DO 1004 K=1,N
1004     D=D-LM(I,K)*AM(K,J)
```

(continued)

Table 6-14 Continued

```
26   1005 DM(I,J)=D
27   1006 DM(I,I)=DM(I,I)+E
28        DO 1007 I=1,N
29        DO 1007 J=1,N
30   1007 LM(I,J)=DM(I,J)
31        DO 1010 I=1,M
32        DO 1010 J=1,N
33        D=0.D0
34        DO 1008 K=1,N
35   1008 D=D-XM(I,K)*AM(K,J)
36        DO 1009 K=1,M
37   1009 D=D+NM(I,K)*CM(K,J)
38   1010 DM(I,J)=D
39        IF (L .GT. M) GO TO 1021
40        DO 1011 I=1,M
41        DO 1011 J=1,N
42   1011 XM(I,J)=DM(I,J)
43        DO 1014 I=1,M
44        DO 1013 J=1,M
45        D=0.D0
46        DO 1012 K=1,M
47   1012 D=D+NM(I,K)*BM(K,J)
48   1013 DM(I,J)=D
49   1014 DM(I,I)=DM(I,I)+E
50        DO 1015 I=1,M
51        DO 1015 J=1,M
```

```
52    1015  NM(I,J)=DM(I,J)
53          E=0.D0
54          DO 1018 I=1,M
55          DO 1017 J=1,M
56          D=0.D0
57          DO 1016 K=1,M
58    1016  D=D+MM(I,K)*BM(K,J)
59    1017  DM(I,J)=D
60    1018  E=E+DM(I,I)
61          F=-E/L
62          DO 1020 I=1,M
63          DO 1019 J=1,M
64    1019  MM(I,J)=DM(I,J)
65    1020  MM(I,I)=MM(I,I)+F
66          GO TO 1003
67    1021  CALL MI(MN,N,LM)
68          DO 1023 I=1,M
69          DO 1023 J=1,N
70          D=0.D0
71          DO 1022 K=1,N
72    1022  D=D+DM(I,K)*LM(K,J)
73    1023  XM(I,J)=D
74          RETURN
75          END
```

Table 6-15
Listing of LIN1A

```
1            SUBROUTINE LIN1(MT,NT,M,N,AM,BM,CM,XM)
2     C     SOLUTION OF XM*AM+BM*XM+CM=0 BY SMITH'S METHOD
3     C     REAL*8 AM(N,N),BM(M,M),CM(M,N),XM(M,N),DM(MN,MN),
4     C    1EM(MN,MN),FM(M,N),MM(M,M),NM(N,N)
5           REAL*8 AM(NT,NT),BM(MT,MT),CM(MT,MT),XM(MT,NT),DM(7,7),
6          1EM(7,7),FM(7,7),MM(7,7),NM(7,7),D,E,RAD,RATIO,DABS
7           LOGICAL*1 H
8           DATA MN/7/,RAD/1.D0/,RATIO/1.D-12/,LMAX/20/
9           DO 1001 I=1,N
10          DO 1000 J=1,N
11          DM(I,J)=-AM(I,J)
12    1000  EM(I,J)=AM(I,J)
13          DM(I,I)=DM(I,I)+RAD
14    1001  EM(I,I)=EM(I,I)+RAD
15          CALL MI(MN,N,DM)
16          DO 1005 J=1,N
17          DO 1003 I=1,M
18          D=0.D0
19          DO 1002 K=1,N
20    1002  D=D+CM(I,K)*DM(K,J)
21    1003  FM(I,J)=D
22          DO 1005 I=1,N
23          D=0.D0
24          DO 1004 K=1,N
25    1004  D=D+EM(I,K)*DM(K,J)
```

```
26  1005  NM(I,J)=D
27        DO 1007 I=1,M
28        DO 1006 J=1,M
29        DM(I,J)=-BM(I,J)
30  1006  EM(I,J)=BM(I,J)
31        DM(I,I)=DM(I,I)+RAD
32  1007  EM(I,I)=EM(I,I)+RAD
33        CALL MI(MN,M,DM)
34        E=RAD+RAD
35        DO 1011 I=1,M
36        DO 1009 J=1,N
37        D=0.D0
38        DO 1008 K=1,M
39  1008  D=D+DM(I,K)*FM(K,J)
40  1009  XM(I,J)=E*D
41        DO 1011 J=1,M
42        D=0.D0
43        DO 1010 K=1,M
44  1010  D=D+DM(I,K)*EM(K,J)
45  1011  MM(I,J)=D
46        L=0
47  1012  DO 1014 I=1,M
48        DO 1014 J=1,N
49        D=0.D0
50        DO 1013 K=1,N
51  1013  D=D+XM(I,K)*NM(K,J)
52  1014  DM(I,J)=D
```

(continued)

Table 6-15 Continued

```
53            H=.TRUE.
54            DO 1016 I=1,M
55            DO 1016 J=1,N
56            D=0.D0
57            DO 1015 K=1,M
58     1015   D=D+MM(I,K)*DM(K,J)
59            IF (DABS(D) .LE. RATIO*DABS(XM(I,J)).) GO TO 1016
60            H=.FALSE.
61     1016   XM(I,J)=XM(I,J)+D
62            IF (H) GO TO 1024
63            L=L+1
64            IF (L-LMAX) 1017,1024,1024
65     1017   DO 1018 I=1,N
66            DO 1018 J=1,N
67     1018   DM(I,J)=NM(I,J)
68            DO 1020 I=1,N
69            DO 1020 J=1,N
70            D=0.D0
71            DO 1019 K=1,N
72     1019   D=D+DM(I,K)*DM(K,J)
73     1020   NM(I,J)=D
74            DO 1021 I=1,M
75            DO 1021 J=1,M
76     1021   EM(I,J)=MM(I,J)
77            DO 1023 I=1,M
```

```
78            DO 1023  J=1,M
79            D=0.D0
80            DO 1022  K=1,M
81      1022  D=D+EM(I,K)*EM(K,J)
82      1023  MM(I,J)=D
83            GO TO 1012
84      1024  RETURN
85            END
```

Table 6-16
Listing of LIN2

```
 1         SUBROUTINE LIN2(MT,MX,NX,RM,CM,ZM)
 2   C     SOLUTION OF XM*AM*AM+AM*XM+BM**XM*BM+CM=0
 3   C     REAL*8 CM(NX,NX),ZM(NX,NX),RM(NT,NT),XV(NT),YV(NT),D
 4         REAL*8 CM(MX,MX),ZM(MX,MX),RM(MT,MT),D
 5         REAL*8 XV(28),YV(28)
 6         NT=(NX*(NX+1))/2
 7         L=0
 8         DO 1000 J=1,NX
 9         DO 1000 I=1,J
10         L=L+1
11         YV(L)=CM(I,J)
12         IF (I .EQ. J) GO TO 1000
13         YV(L)=2.D0*YV(L)
14  1000   CONTINUE
15         DO 1002 I=1,NT
16         D=0.D0
17         DO 1001 J=1,NT
18  1001   D=D-RM(I,J)*YV(J)
19  1002   XV(I)=D
20         I=0
21         DO 1003 J=1,NX
22         DO 1003 I=1,J
23         L=L+1
24         ZM(I,J)=XV(L)
25  1003   ZM(J,I)=XV(L)
26         RETURN
27         END
```

Subprogram INC, which is used to compute equivalent coefficient matrix Δ and its inverse, is summarized as follows:

INC function: Compute the inverse of the equivalent coefficient matrix of

$$XA + A'X + \sum_{k=1}^{m} B_k' X B_k$$

as a subprogram.

INC dummy argument list: MT,MX,MV,NX,NV,AM,BA,ZM, where
$1 \leqslant MT \leqslant 28, 1 \leqslant MX \leqslant 7, 1 \leqslant MV \leqslant 3, 1 \leqslant NX \leqslant MX, 0 \leqslant NV \leqslant MV$.
AM(MX,MX) = A.
BA(MV,MX,MX) = $B_1/B_2/\ldots$.
ZM(MT,MT) = inverse of equivalent coefficient matrix.

INC input data list: None.

INC output data list: None.

Subprograms called by INC: MI.

Subprogram INC also is listed in Table 6-17.

MI and MIA

Subprograms MI and MIA are used to invert matrices and are summarized as follows:

MI and MIA function: Compute the inverse of A as a subprogram.

MI and MIA dummy argument list: M,N,AM, where
$1 \leqslant M \leqslant 30, 1 \leqslant N \leqslant M$.
AM(M,M) = A.

MI and MIA input data list: None.

MI and MIA output data list: None.

Subprograms called by MI and MIA: None.

Subprogram MI is listed in Table 6-18; subprogram MIA, in Table 6-19

Table 6-17
Listing of INC

```
1            SUBROUTINE INC(MT,MX,MV,NX,NV,AM,BA,ZM)
2     C      INVERSE OF REDUCED EQUIVALENT COEFFICIENT MATRIX
3     C      REAL*8 AM(NX,NX),BA(NV,NX,NX),ZM(NT,NT)
4     C      INTEGER*4 IM(NX,NX)
5            REAL*8 AM(MX,MX),BA(MV,MX,MX),ZM(MT,MT)
6            INTEGER*4 IM(7,7)
7            NT=(NX*(NX+1))/2
8            DO 1000 I=1,NT
9            DO 1000 J=1,NT
10    1000   ZM(I,J)=0.D0
11           L=0
12           DO 1001 J=1,NX
13           DO 1001 I=1,J
14           L=L+1
15           IM(I,J)=L
16    1001   IM(J,I)=L
17           L=0
18           DO 1006 J=1,NX
19           DO 1006 I=1,J
20           L=L+1
21           DO 1004 K=1,NX
22           M=IM(I,K)
23           N=IM(K,J)
24           ZM(L,M)=ZM(L,M)+AM(K,J)
25           ZM(L,N)=ZM(L,N)+AM(K,I)
26           IF (NV)  1004,1004,1002
27    1002   DO 1003 M=1,NV
28           DO 1003 N=1,NX
29           NP=IM(K,N)
30    1003   ZM(L,NP)=ZM(L,NP)+BA(M,K,I)*BA(M,N,J)
31    1004   CONTINUE
32           IF (I .EQ. J) GO TO 1006
33           DO 1005 K=1,NT
34    1005   ZM(L,K)=2.D0*ZM(L,K)
35    1006   CONTINUE
36           CALL MI(MT,NT,ZM)
37           RETURN
38           END
```

Table 6-18
Listing of MI

```
 1            SUBROUTINE MI(M,N,AM)
 2     C      INVERSE OF AM BY THE PIVOT METHOD
 3     C      REAL*8 AM(N,N),DM(N,N),DV(N),EV(N),D,E,DABS
 4     C      INTEGER*4 FV(N),GV(N)
 5     C      LOGICAL*1 HV(N)
 6            REAL*8 AM(M,M)
 7            REAL*8 DM(30,30),DV(30),EV(30),D,E,DABS
 8            INTEGER*4 FV(30),GV(30)
 9            LOGICAL*1 HV(30)
10      1000  FORMAT(' 1')
11      1001  FORMAT(//,'      SINGULAR MATRIX')
12            DO 1002 I=1,N
13            HV(I)=.FALSE.
14            DO 1002 J=1,N
15      1002  DM(I,J)=AM(I,J)
16            DO 1008 K=1,N
17            L=0
18            D=0.D0
19            DO 1003 J=1,N
20            IF (HV(J)) GO TO 1003
21            E=DABS(DM(K,J))
22            IF (E .LE. D) GO TO 1003
23            D=E
24            L=J
25      1003  CONTINUE
26            IF (L) 1004,1004,1005
27      1004  WRITE (6,1001)
28            WRITE (6,1000)
29            RETURN
30      1005  D=1.D0/DM(K,L)
31            DO 1006 I=1,N
32            DV(I)=D*DM(I,L)
33            EV(I)=DM(K,I)
34            DM(I,L)=0.D0
35      1006  DM(K,I)=0.D0
36            DV(K)=D
37            EV(L)=-1.D0
38            DO 1007 I=1,N
39            DO 1007 J=1,N
40      1007  DM(I,J)=DM(I,J)-DV(I)*EV(J)
```

Table 6-18 Continued

```
41            DM(K,L)=D
42            FV(K)=L
43            GV(L)=K
44      1008  HV(L)=.TRUE.
45            DO 1009 I=1,N
46            K=FV(I)
47            DO 1009 J=1,N
48            L=GV(J)
49      1009  AM(K,L)=DM(I,J)
50            RETURN
51            END
```

Table 6-19
Listing of MIA

```
1            SUBROUTINE MI(M,N,ZM)
2      C     POS. DEF. MATRIX INVERSION BY SQUARE ROOT DECOMPOSITION
3      C     REAL*8 ZM(M,M),ZIM(M,M)
4            REAL*8 ZM(M,M),ZIM(30,30),D,DD,DY,DSQRT
5            NA=N-1
6            DO 1007 I=1,N
7            D=ZM(I,I)
8            NN=I-1
9            MM=I+1
10           IF (NN) 1002,1002,1000
11     1000  DO 1001 J=1,NN
12     1001  D=D-ZIM(J,I)**2
13     1002  DY=1.D0/DSQRT(D)
14           IF (N-MM) 1007,1003,1003
15     1003  DO 1006 J=MM,N
16           D=ZM(I,J)
17           IF (NN) 1006,1006,1004
18     1004  DO 1005 K=1,NN
19     1005  D=D-ZIM(K,I)*ZIM(K,J)
20     1006  ZIM(I,J)=DY*D
21     1007  ZIM(I,I)=DY
22           IF (N-2) 1010,1008,1008
23     1008  DO 1009 I=2,N
24           NN=I-1
25           DO 1009 J=1,NN
```

(continued)

Table 6-19 Continued

```
26       1009 ZIM(J,I)=ZIM(I,I)*ZIM(J,I)
27       1010 IF (NA) 1017,1017,1011
28       1011 DO 1016 I=1,NA
29            NN=I-1
30            MM=I+1
31            IF (NN) 1016,1016,1012
32       1012 DO 1015 J=1,NN
33            L=J+1
34            D=-ZIM(J,J)*ZIM(J,MM)
35            IF (I-L) 1015,1013,1013
36       1013 DO 1014 K=L,I
37       1014 D=D-ZIM(J,K)*ZIM(K,MM)
38       1015 ZIM(J,MM)=D
39       1016 ZIM(I,MM)=-ZIM(I,I)*ZIM(I,MM)
40       1017 DO 1021 I=1,N
41            DO 1021 J=I,N
42            L=J+1
43            D=ZIM(I,J)*ZIM(J,J)
44            IF (N-L) 1020,1018,1018
45       1018 DO 1019 K=L,N
46       1019 D=D+ZIM(I,K)*ZIM(J,K)
47       1020 ZM(I,J)=D
48       1021 ZM(J,I)=D
49            RETURN
50            END
```

References

[1] Merriam, C.W., III: *Automated Design of Control Systems,* Gordon and Breach, Science Publishers, New York, 1974.
[2] Bryson, A.E., Jr., and Ho, Y.C.: *Applied Optimal Control,* Blaisdell Publishing Co., Waltham, 1969.
[3] IBM System/360 and System/370 FORTRAN IV Language, GC28-6515-10, International Business Machines Corp., New York, 1974.
[4] Kuester, J.L., and Mize, J.H.: *Optimization Techniques with FORTRAN,* McGraw-Hill, New York, 1973.
[5] Merriam, C.W., III: Improved second-order methods for solving constrained parameter optimization problems arising in control, *Int. J. Control,* 1972, 15:(5)865-876.
[6] Jordan, D., and Merriam, C.W., III: Synthesis of transfer functions matrices in control. *Prod. 1972 IFAC Congr.* Paris, 1972.
[7] Faddeeva, V.N.: *Computational Methods of Linear Algebra,* Dover Publications, New York, 1959.
[8] Gill, S.: A process of step-by-step integration of differential equations in an automatic digital computing machine. *Proc. Cambridge Phil. Soc.* 47 (Pt. 1), June, 1950.
[9] Merriam, C.W., III: A computational method for feedback control optimization. *Inform. Contr.,* 8(2): 1965.
[10] Merriam, C.W., III: *Optimization Theory and the Design of Feedback Control Systems,* McGraw-Hill, New York, 1964.
[11] Ralston, A.: *A First Course in Numerical Analysis,* McGraw-Hill, New York, 1965.
[12] Smith, R.A.: Matrix Equation $XA + BX = C$. *Siam J. Appl. Math.* 16(1): 1968.

Index

Asymptotic stability constraint, 214, 248

Conjugate gradient, 9, 10, 13, 21, 114
CONVEX, 2, 3, 81, 286, 294, 301, 306, 310, 321

Direction vector, 9, 10, 42, 43, 176, 184
DXM, 2, 3, 101, 145, 150

Effective inequality constraints, 42
Equality constraints, 25, 26, 44, 213
EVAL, 2, 3, 81, 286, 294, 301, 306, 311, 314, 321
EVL, 1, 3, 81, 82, 83
Excess pole specification, 69, 70, 72, 91

FACT, 2, 3, 81, 285, 286, 293, 294, 295, 314, 321
FACTOR, 2, 3, 81, 285, 286, 287, 314, 320
Fletcher-Powell method, 43, 44
Forcing-function optimization, 167, 176, 213
FORM1, 1, 3, 92, 93, 94, 98
FORM2, 1, 3, 127, 128, 129
FUN, 2, 3, 184, 185, 191

GAIN1, 1, 3, 92, 101, 110, 111, 115
GAIN2, 1, 3, 138, 139, 143
Gill modification, 144, 168
Gradient, 9, 92, 110, 144, 167, 168, 184, 282, 311
Greatest common divisor polynomial, 282, 286, 315

Hamiltonian, 167
Hessian matrix, 11, 25, 282, 311, 323
Horner's method, 314

INC, 2, 3, 101, 153, 332, 341, 342
Inequality constraints, 25, 26, 37, 43, 44
INT, 1, 3, 144, 146, 157
INTC, 4
INTL, 2, 3, 81, 286, 294, 300, 302, 306, 321
INTS, 4, 145, 152, 157
Iteration, 9, 10, 26

Jacobian matrix, 25

Lagrange multiplier, 26, 42, 267
Lagrangian, 25, 42, 43
Lehmer-Schur test, 283, 294, 300, 305, 320

Linear optimal control, 247
LIN1, 2, 3, 110, 330, 331, 333
LIN1A, 2, 3, 330, 331, 336
LIN2, 2, 3, 101, 153, 332, 340
LOAD1, 2, 3, 92, 110, 114, 116
LOAD2, 2, 3, 152, 153, 160
LOAD3, 2, 3, 168, 175, 179
LOAD3S, 4, 169, 175, 176
Lyapunov equations, 92, 142, 323

MATCH, 2, 3, 81, 286, 315, 316, 321
MI, 2, 3, 27, 44, 71, 93, 101, 110, 128, 138, 153, 216, 249, 330, 341, 343
MIA, 2, 3, 331, 341, 345
Min-max method, 42, 43, 44, 61
MIN1, 1, 3, 13, 14, 22, 92, 110, 123, 144, 152, 153, 165, 168, 176, 179, 184
MIN1C, 4, 21, 114, 153, 176
MIN1S, 4, 13, 21, 110, 114, 153
MIN2, 1, 3, 26, 28, 37, 38
MIN2C, 4, 37
MIN2S, 4, 27, 37
MIN3, 1, 3, 44, 45, 61, 92, 110, 144, 152, 153, 169, 184
MIN3C, 4, 61, 114, 153
MIN3S, 4, 44, 61, 110, 114, 153
MIN4, 1, 3, 184, 185, 191, 192, 207
MIN4C, 4, 192
MIN4S, 4, 184, 185, 191, 192
MIN5, 3, 215, 216, 217, 236
MIN5C, 4, 236
MIN5S, 4, 216, 236

Newton method, 92, 128, 142
Newton-Raphson method, 25–26, 37, 42–43, 44, 61, 213, 215, 247, 282–284, 294, 306, 311, 320, 323

Objective function, 9, 25, 42, 43, 92, 110, 127, 144, 152, 153, 167, 184, 281, 282
Optimal gain control, 91, 92, 101, 110, 176

Parameter optimization, 167, 176, 192, 213, 281
Performance functional, 167
Performance integral, 70, 91
PC, 2, 3, 81, 286, 315, 317, 318, 321
PD, 2, 3, 81, 286, 315, 316, 319, 321
Pivot method, 330
PM, 2, 3, 318, 320, 321

QUAD, 1, 3, 92, 93, 98, 102, 107, 127, 128, 137

Quadratic convergence, 37, 101, 236, 323

Riccati equations, 92, 98, 127, 214
Runge-Kutta, 144, 168

SEARCH, 2, 3, 81, 286, 293, 294, 306, 311, 312, 321
SIM, 1, 3, 248
SIMC, 4, 267
SIMS, 4, 249, 267
Slack variables, 25–26, 42
Square root decomposition, 331

Steepest descent, 9, 10, 13, 169, 185
Successive approximations, 13, 44, 144, 184, 185
Successive substitutions, 26, 92, 214, 215
SYN, 1, 3, 71

TEST, 1, 3, 320, 325
Trial, 11, 12, 13, 84
Two-point boundary-value problem, 168, 176, 247

ZEROS, 2, 3, 81, 294, 301, 305, 307, 321

About the Author

Charles Merriam has held a number of academic positions at M.I.T., Cornell, and the University of Rochester, where he currently is professor and chairman of the Department of Electrical Engineering. He has also held a number of research and consultant positions with the General Electric Company where most of the computational techniques reported here were developed. He is the author of *Optimization Theory and the Design of Feedback Control Systems,* a pioneering work that has been translated into both Russian and Polish; *Analysis of Lumped Electrical Systems;* and *Automated Design of Control Systems,* which includes much of the theoretical foundation for the computational techniques presented in this book.